W9-ATH-766

QH
541.5
.S3
S66
DISCARDED
JENKS LRC
GORDON COLLEGE

The Structure of Marine Ecosystems

John H. Steele

Harvard University Press
Cambridge, Massachusetts
and London, England

Second printing 1975
Library of Congress Catalog Card Number 73-82350
ISBN 0-674-84420-3
Printed in the United States of America

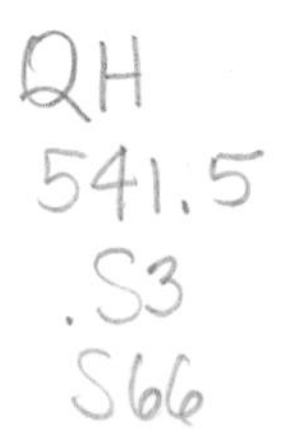

The Structure of Marine Ecosystems

Acknowledgments

I should like to dedicate this short work to Cyril Lucas, whose guidance over many years provided the basis for development of the ideas I have attempted to express here.

Roger Bailey and Michael Mullin made detailed and helpful comments on the original draft of the manuscript, and many of my colleagues in Aberdeen provided valuable advice and assistance. To all of them, I express my deep gratitude.

Thanks are also due the editors and authors of various books and scientific papers for their permission to use illustrations and other material; in particular, I am grateful to Almqvist and Wiksell, Stockholm, for material from my paper, "The Changing Chemistry of the Oceans," in *Nobel Symposium 20* (Dyrssen and Jagner, eds.).

Contents

Tables

Figures

Figures

The Structure of Marine Ecosystems

1

Introduction

Natural systems can be studied in several ways, and the methods used tend to emphasize differing types of patterns. The intrinsic fascination with the great variety of species present in any area of land or water can be developed into the quantitative, usually statistical, analysis of community structure, displaying similarities in relative abundance of species within different communities and defining concepts such as "niches." At the other extreme, the intricacies of social or sexual behavior of a particular species can be studied in continuously increasing detail to reveal the complexity and variety of possible adaptations to environment. The general structure of communities and the behavior of an individual when feeding, breeding, or escaping from predators provide, indirectly, some idea of how species interact with one another. Between these two approaches is the study of energy flow through food webs, where links between species—or groups of species—is the primary concern. This method appears relatively neutral, yet it contains implicitly the assumption that energy is the unit by which other variables can be quantified. The overtones of the term "efficiency," when used to describe the growth of individuals or to compare communities, emphasize the presumed generality.

All these methods have one common problem. We wish to know not only how many species there are, how each individual behaves, or how much energy it exchanges with the rest of the food web; we hope to discover why, in any ecosystem, there is a particular distribution of species and of energy flow and how this distribution persists over periods of time. This leads to questions about the stability of communities and about density dependence in the responses between the components of each community. The answers are unlikely to come from any single method of study. There have been attempts to bring these various aspects together to form unified pictures of reality, but often these are too simple and too general to be of help in the analysis of an individual situation. Since any theory must start and end with close attention to field data and experiments, the types of approach which can be used are limited thereby.

In the present study of marine problems I have found it necessary to choose one viewpoint and have assumed that the rates of energy flow in an ecosystem are the most significant parameters. This choice implies the ability to express in terms of calories the rates at which individual organisms assimilate food, metabolize and excrete part of this food, grow, reproduce, and finally die or are eaten by other animals. It does not mean that calories are the sole unit used. The rate of supply of nitrogen for protein synthesis may be a limiting factor, but it should be possible to relate it to the rates of change of energy in a given population. The construction of energy flow charts is a necessary step, but it does not provide answers to questions about the size of populations. These answers appear to depend on a more detailed understanding of population interactions. I assume that analyzing the behavior of individuals in the population is the most useful way of studying this aspect.

Details of behavior—such as production of and response to chemical stimuli—would be very difficult to quantify in energetic terms, beyond noting that the energy requirements of these processes in general are very small. Implicit in the use of energy as a unit is the assumption that for any organism the specific methods used to detect a possible item of food or a potential mate have evolved to an extent that the energy required is minimized. It is the consequences of the eating and mating responses that are most significant in the energy budgets of the organism and the population. For example, changes in rate of predation as a function of variations in density and species of prey will be discussed here in terms of several very general types of response, which undoubtedly operate through detailed behavioral patterns of which we usually have little or no knowledge. Thus an arbitrary and simple definition of behavior is used here.

These assumptions are easier to believe in marine ecology, where normally one cannot observe the organisms directly. The relative uniformity of seawater as an environment, when compared with land, is probably an illusion, although the dominance of certain groups (such as herbivorous copepods) does not seem to have a parallel in terrestrial systems.

One of the difficulties is that the types of investigation performed at sea and on land are usually very different. Whereas a field or a wood, or even a single tree, may be the unit of study on land, the areas in the open sea generally are much larger—for example, the North Sea or Long Island Sound. These larger units are considered necessary because of the continuous intermixing of the water by currents and by turbulent diffusion. The traditional procedure is for a ship to steam along transects, sampling the water and the plant and animal life every few miles. One can try to avoid this by investigating a single position in the open sea and assuming that conditions are uniform over a sufficiently large area around this position. Since we know that the plankton which move with the water are normally patchy in their distribution even when there are no observable differences in the physical environment, this assumption applies only within certain limits and would not usually be acceptable on land. Yet it is this mixing of water masses which permits us the concept of more even distributions in the sea than on land.

Some of the questions that emerge from these considerations are, How much plant production is there in an upwelling region such as that off Peru, and how do physical processes of nutrient enrichment control it? Can changes in the plankton species composition across the North Atlantic be related to changing water masses? How does one explain the large fluctuations in yearly recruitment to the commercially exploited fish stocks in the North Sea? By asking questions like these we imply the existence of fairly broad hypotheses relating populations to their environment. It may be the nature of the sampling methods rather than the assumed simplicity of the marine environment which makes us less inhibited in using simple models to reach general conclusions. Because we cannot move freely within the deep sea, it is desirable to test our hypotheses by comparison with relevant studies on land. This may appear unreasonable when some of the differences are considered (the abundance of grass and trees as opposed to the scarcity of phytoplankton, for instance). Yet I think there is validity in asking whether there are any similarities between mechanisms proposed for the long-term stability of terrestrial and of marine ecosystems.

In a provocative paper Hairston, Smith, and Slobodkin (1960) gave their opinion on terrestrial systems:

> All organisms taken together are limited by the amount of energy fixed. In particular the decomposers as a group must be food limited . . . Producers [plants] are neither herbivore-limited nor catastrophe-limited and must therefore be limited by their own exhaustion of a resource . . . It therefore follows that the usual condition is for populations of herbivores *not* to be limited by their food supply. . . Although rigorous proof that herbivores are generally controlled by predation is lacking, supporting evidence is available and the alternative hypothesis of control by weather leads to false or untenable implications . . . The predators and parasites, in controlling the population of herbivores, must thereby limit their own resources, and as a group they must be food limited . . . Interspecific competition exists among producers, among carnivores, and among decomposers.

This opinion is echoed by Lack (1966):

> Though I suggested that the numbers of most birds, carnivorous mammals, certain rodents, large fish where not fished, and a few insects are limited by food, I suggested that the numbers of gallinaceous birds, deer and phytophagous insects for at least most of the time are limited by predators (including insect parasites); and it may be added that phytophagous insects comprise the great majority of the world's animal species.

Recently, in relation to pollution problems, the following assessment was made (MIT, 1970):

> Terrestrial ecosystems of landscape size usually contain hundreds of species of plants and thousands of species of animals, but the herbivores

rarely consume more than 5 to 15 per cent of the vegetation. The normal degree of control is best summarized by pointing out that for each species of terrestrial plant there are about 100 species of animals capable of eating it, yet most of the time, most of the plant production falls to the ground uneaten. For many insects able to defoliate trees the population density rarely rises as high as one per tree. Nevertheless, this stability is fragile enough to be easily destroyed. The list of disturbances that lead to population outbreaks (almost always of herbivores) includes almost anything that removes a number of species or impairs the health or numbers of predators.

All of these conclusions depend on the fact that, except in managed systems, less than 10 percent of plant material is eaten while it is still living, so that nearly all of the energy produced photosynthetically goes to the decomposer cycle. The marine system is quite different. The phytoplankton of the open sea is eaten nearly as fast as it is produced, so that effectively all plant production goes through the herbivores. The animals living on the sea bottom depend on herbivore feces, rather than on a direct fallout of plants, for their food supply. One of the main technical problems in plankton studies has been to demonstrate experimentally that the herbivorous zooplankton can get enough to eat from the densities of phytoplankton found normally in the sea. All indications suggest that herbivores in the sea *are* resource-limited. On the other hand, there is evidence that these herbivores are highly efficient at transferring energy through the food chain from plants to primary carnivores. It seems inadequate to suggest that marine herbivores, operating efficiently just above starvation level, achieve a balance between their numbers and their food supply solely through control by carnivores.

Theoretical studies by Holling (1965) have emphasized the dichotomy between "simple-minded" invertebrates, which search for food at the same rate for all food concentrations, and the more complex response of vertebrates, which are considered to have a threshold for a particular food item below which their feeding rate is greatly decreased. Holling and others point out that the latter form of behavior can have a stabilizing effect on the ecosystem. Although it is not directly comparable, this theory has analogies with the hypothesis of

Hairston and his associates that herbivores would decimate their food unless controlled in a density-dependent manner by the carnivores. From either point of view, it seems unlikely that small invertebrate herbivores such as copepods could have responses to varying food concentrations of the type necessary to determine their own population stability. Yet there is now sufficient evidence to indicate that this does happen in the sea and that there is a threshold response of copepods to changes in food concentration similar to that proposed for terrestrial carnivores. Thus there may be analogies between terrestrial and marine systems, but these do not arise from a direct comparison of plants, herbivores, and carnivores in the two systems. (If one is sufficiently imaginative, trees might appear to be the nearest analogue to fish. Both produce very large numbers of seeds, or eggs, which have very high initial mortalities followed by very low death rates until man is involved in their harvest.)

These points will be considered in detail later on, but the main problem in presentation is to display the consequences of combining the various hypotheses about feeding behavior at different trophic levels. Since I am assuming that the hypotheses can be stated in quantifiable terms related to food and metabolism, it follows that they can be expressed as mathematical formulas and then combined into models. Such models do not contain any new facts, but they can show the results of our assumptions—for example, by illustrating the effects of different rates of nutrient uptake by plants on the numbers and growth rates of herbivores.

As the model used here is developed, it will be found that a large number of parameters is needed to describe even the essential features of a single plant-herbivore-carnivore interaction. This model could be made more "realistic" by including other variables, such as thermocline depth or solar radiation, and by giving these and other parameters a stochastic component. Such additions would not only make the model more complicated but would make it more difficult to assess the effects of variations in, say, grazing rate or respiration rate of copepods. Thus the model is kept as simple as possible and is an idealized picture of spring and summer in higher latitudes. The actual numbers used are closest to events in the northern North Sea, the area of which I have most knowledge. In this way the model will provide illustrations for the text, rather than proofs of the main line of the argument. As far as possible the various hypotheses and results will be stated verbally or graphically.

This, or any, model needs to be set in the context of field observations and so some examples of marine food webs are given in Chapter 2. Although they are intended as illustrations, they have been chosen to demonstrate the relatively high efficiency with which energy can be transferred from plants to carnivores through the herbivores—one of the main differences between marine and terrestrial systems. It is sometimes considered that efficiency and stability are incompatible, the former associated with a low, and the latter with a high, diversity of species in a community. Part of the problem here is semantic, arising particularly from the definition of stability. I have chosen to approach this mathematically (in Chapter 3), extending the simple two-species Lotka-Volterra model of prey-predator interaction to the case of a large number of species at each of several trophic levels. If all the reactions between species are of the simple linear form used in the Lotka-Volterra equations, there is still the same lack of positive stability associated with the two-species system. In other words, diversity in itself does not induce stability, which must arise through more complicated forms of behavior than simple linear responses to changes in prey or predator density. These are discussed in more detail in Chapter 3 and, with reservations, this argument is used to justify a food *chain* simulation with complex responses rather than an attempt to simulate a *web*.

The model of a marine ecosystem in Chapter 5 brings out not only the necessity for a threshold-type response at the herbivore level, but also the fact that similar responses may occur at higher levels and will improve the ability of the system to cope with environmental fluctuations (Chapter 6). The general, theoretical conclusion of Chapters 3 and 5 is that there is no simple separation of factors controlling stability, with control found solely in the herbivore-carnivore interactions on land and in the plant-herbivore interactions in the sea. Even so, we shall still deduce that a breakdown in control at these particular points in each system has much more severe consequences for the whole system than changes at other points. Thus the differences in structure of marine and terrestrial systems do imply differences in points at which control is critical; but it appears possible that the similarities in functional response, even though these are at different trophic levels, may permit certain types of generalizations which could include both systems.

Such generalizations depend in part on the degree to which our model can portray real situations. The problem with this, or with any, model is to find data for testing it. The difficulties of making suitable observations at sea are discussed in Chapter 4, and some of the results of detailed surveys are used to show the adequacy or inadequacy of the model. Predictions resulting from the model led to experimental work on zooplankton respiration, which is described in Chapter 6.

The main aim of this analysis of marine ecosystems is to show how theory, observation, and experiment may be combined, and how closely each depends on the other. It also provides a basis for speculation about the effects of man's intervention in marine ecosystems and the way in which these differ from the consequences of his action on land (Chapter 7). Our excessive fishing activities can greatly alter or destroy certain fisheries without any catastrophic changes in the rest of the system. On the other hand, eutrophication in fresh water, by altering the plant populations, appears to induce changes in the whole of the system. These examples can be fitted into the theoretical framework, and this framework can be used to approach recent or future problems of man's impact in the open sea.

2

Marine Food Webs

There is abundant evidence of the complexity of food webs in the sea. Hardy's (1924) description of the feeding relations of the herring provides the classic illustration. There is not a simple chain from phytoplankton through herbivorous copepods to herring, since other predators on the copepods (such as the arrowworm *Sagitta* and the sand eel *Ammodytes*) also are eaten by herring. The effect of such interaction is to decrease the potential yield that could be taken by a carnivore at a higher trophic level (see Fig. 2.1).

It needs to be emphasized that this is not merely a consequence of species diversity. Ten carnivorous species preying on a hundred herbivore species, which in turn are grazing on a thousand plant species, is certainly more diverse than a chain with only one species of carnivore, herbivore, and plant. It may be more stable, but there is no intrinsic reason why the former should yield less carnivore biomass than the latter. It is the number of "horizontal" links in the food web which is important, since it is these which degrade energy within a trophic level. Such links occur when an animal feeds at more than one trophic level, as in Fig. 2.1, and such animals are very common in the sea. For example, adult demersal fish such as whiting and cod feed to some extent on other fish and so are secondary carnivores, while their larvae, feeding on copepods, are primary carnivores. A further factor is the time scale. Sand eels, which live for at least two years, provide a source of food which will fluctuate less markedly than copepods, with a life span of about two months.

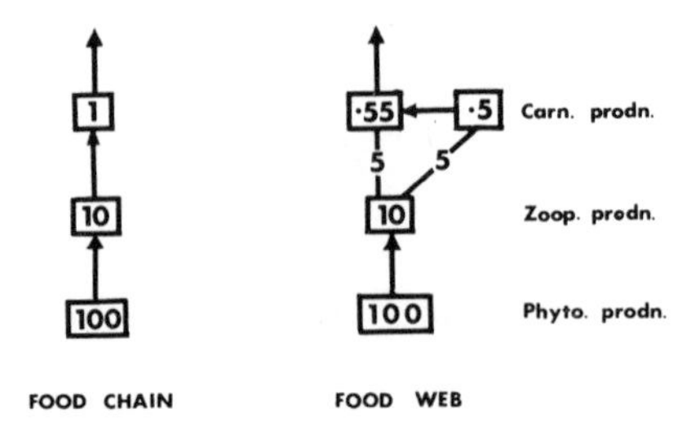

FIGURE 2.1 A simple example of the decreased energy output caused by branching in a food chain, assuming 10 percent efficiency in transfer through each production level.

The Black Sea

Unfortunately there are very few studies combining information on feeding patterns in terms of species with estimates of the energy transfers and losses at each of these links in the food web. A detailed investigation of the food web found in winter in the Black Sea (Petipa, Pavlova, and Mironov, 1970) provides some information on both aspects. They consider two "communities"—the epiplankton above the thermocline (16° to 17° C) and the bathyplankton below it (7° to 12° C). Many species migrate between the two zones, so that the faunal lists are not so different; however, they are separated, since the data give different ratios for transfer efficiency, defined as the ratio of calories eaten from a given trophic level to calories consumed by that trophic level. The trophic levels (Table 2.1) are based on a detailed study of feeding habits.

Thus, although certain copepods are purely phytoplankton feeders (herbivores), others also feed at times on the naupliar stages of their own or other species (mixed consumers), while one level (primary carnivores) consists of adult copepods which in early life are members of other trophic levels. The arrowworms *Sagitta* are defined as secondary carnivores because some of their food items are carnivorous copepods. Similarly, *Pleurobrachia* (a Ctenophore) and Medusae are tertiary because they sometimes eat *Sagitta* as well as copepods. (Their transfer efficiency in the epiplankton was calculated using population changes.)

TABLE 2.1 Efficiency of different trophic levels in the Black Sea (Petipa, Pavlova, and Mironov, 1970).

	Transfer efficiency (%)	
	Epiplankton	Bathyplankton
Phytoplankton		
Herbivores (copepods such as *Pseudocalanus* and *Calanus*)	71	25
Mixed food consumers (copepods such as *Acartia* and juvenile *Oithona*)	42	36
Primary carnivores (adult *Oithona* sp.)	19	29
Secondary carnivores (*Sagitta setosa*) } Tertiary carnivores (*Pleurobrachia*, Medusae)	7	0

The main feature of the results is the high values for "transfer efficiency." Since these were calculated for only a short part of a year rather than for a whole year, they may be nearer to growth efficiencies than ecological efficiencies,* but even so they are exceptionally high. There are also, as always, technical problems concerning the effectiveness of the nets in catching the smaller plankton, whose omission could affect the results. However, these data do provide a first illustration of the apparently high efficiency with which energy is transferred through the pelagic food chain.

Since these results are much more detailed than other investigations

*Growth efficiency is the ratio of growth to ingestion for an average individual during part of its life. Ecological efficiency is the transfer efficiency of a population averaged over a long enough period for the population to be in steady state.

to be discussed later, it will be helpful to transform them into a different pattern. In Fig. 2.2 the trophic level of a group is determined as one above the lower level at which it feeds. This fits in with our normally less detailed knowledge of feeding patterns. It also follows from the probability, illustrated earlier, that the lower level provides most of the energy. Finally, it enables one to see how many horizontal links there are, and thus what "price" the ecosystem is paying for the possible advantage of particular groups feeding at different levels. From the data available in Petipa one can calculate the ratio of calories eaten by secondary-tertiary consumers to calories eaten by herbivores plus mixed food consumers. The values are 25 and 22 percent for epi- and bathyplankton communities respectively.

The Pacific Sardine

The second example concerns the Pacific sardine, whose energy budget has been worked out in detail by Lasker (1970). Figure 2.3 illustrates the decrease in growth efficiency with age so that for a given quantity of food supply for the fish, the yield from such a fishery is increased by taking the fish as young as possible. Conversely, this decreases the spawning potential and can make the stock more susceptible to other random environmental fluctuations. Between the years 1932–1934 and 1951–1956 the biomass of Pacific sardine decreased to one-seventh of its initial level. According to Murphy (1966, as quoted by Lasker), the reduction in the sardine population resulted from successive years when the sardine spawn did not survive sufficiently, coupled with severe fishing on the spawning stock which crippled the future spawning potential of the population. If one accepts the hypothesis that an excessive harvesting of fish about 3 years old can cause such a decline, then from Fig. 2.3 it would appear that the "efficiency," expressed here as the ratio of "calories yield to calories assimilated" up to any age, must be less than 10 percent to ensure survival of this population by allowing several years for reproduction.

In the same paper, Lasker compares data on primary production supplied by Strickland with the food requirements of the sardines,

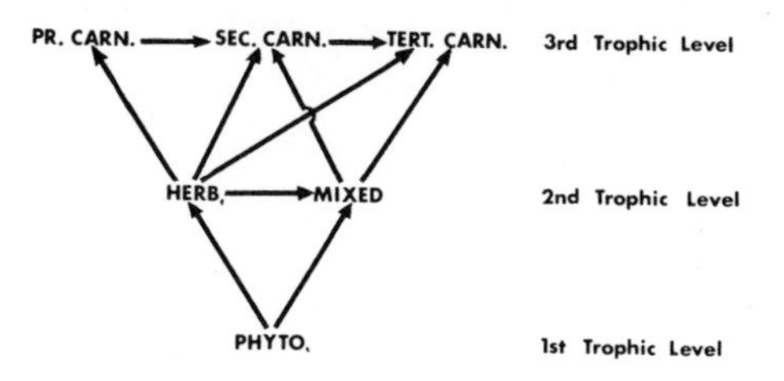

FIGURE 2.2 A simplified version of the food web in the Black Sea (from Petipa, Pavlova, and Mironov, 1970).

which are considered to be predominantly feeders on herbivorous copepods. This ratio—sardine food requirement to primary production—has a value of 0.22 for the main feeding season of the sardines, January to June, during the years 1932–1934 of maximum sardine population. Thus, assuming there were no other predators on the zooplankton, their transfer efficiency is greater than 20 percent. The food chain is certainly not so simple. It is possible that phytoplankton may form part of the food of sardines, but on the other hand it is unlikely that sardines were the only predators on the copepods during the spring and early summer. Although it is circumstantial, this evidence provides the same picture of high herbivore and lower carni-

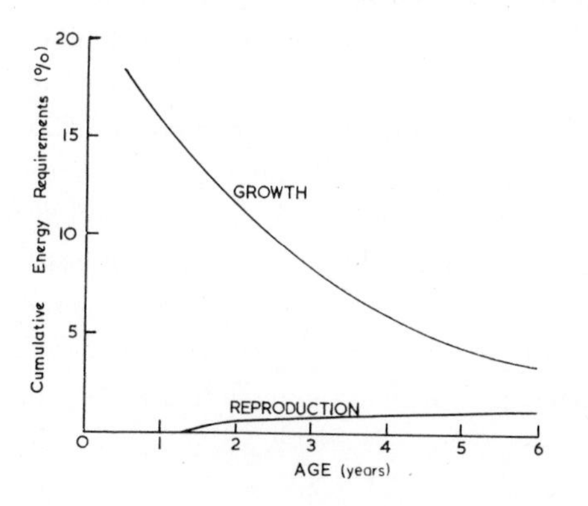

FIGURE 2.3 The cumulative percentage of energy assimilated which is used by the Pacific sardine for growth and reproduction up to any age (from Lasker, 1970).

vore efficiencies that was found in the Black Sea epiplankton of Table 2.1.

Lasker's results illustrate a common feature of our knowledge of marine food chains. We now have considerable data on primary production, mainly from use of the C^{14} technique developed by Steemann Nielsen (1952) to measure carbon assimilation by phytoplankton. At the other extreme, for areas where there is intensive fishing, we have data on the yield to man of commercial fish species. The difficulty is that for the intermediate trophic levels our knowledge of the natural populations is often limited to the biomass rather than to the energetics of the pelagic herbivores and benthic invertebrates which are the food of the commercial fish stocks. The zooplankton are sampled by nets, and the area of the mouth of the net, the mesh size, and the speed of towing can all affect the composition of the catch. The data are usually presented as numbers of each species together with the dry weight of the catch, expressed in cubic meters of water filtered or as numbers and weight under one square meter of surface area. Dominant species of copepods can be identified to their naupliar and copepodite stages, but these data cannot be converted into production estimates unless the length of time spent in each stage is known. Attempts have been made to do this for natural populations of *Calanus finmarchicus* (Marshall and Orr, 1955) but it is not clear how widely these time intervals may vary with temperature and, especially, with food supply. Experimental rearing of *Calanus* sp. with different algal cultures as food, and at different temperatures (Mullin and Brooks, 1970a) show variations in growth rate with both factors. For the benthos, sampled by grabs or corers, determination of the age of dominant components such as polychaetes is not possible directly. Only the longer-lived mollusks with yearly rings on their shells can be used for estimates of growth rate. Thus data from commercial fisheries are invaluable in providing a minimum rate of output from a food web.

The North Sea

The North Sea is one of the most intensively fished areas in the world. It also has the longest records on the yields of fish. I have divided these into two groups, "pelagic" and "demersal," on the basis of what is known of their feeding habits. Using data (published by the International Council for the Exploration of the Sea) for the area shown in Fig. 2.4, Fig. 2.5 shows the effects on the yields of fish of increased fishing intensity combined with the development of more efficient methods of catching (particularly the use of the purse-seine for pelagic species such as herring and mackerel). If we use the data for the five years 1965 to 1969 as an estimate of the maximum fishing yield, we have average values of 2.04×10^6 tons and 0.93×10^6 tons for the annual yield of pelagic and demersal species respectively.

The marked increase in pelagic yield shown in Fig. 2.5 is caused partly by the increase in the catches of mackerel which, for the five years in question, averaged 30 percent of the total pelagic yield compared with 3 percent in earlier years. This increased yield results in part from the capture of older fish from a previously unfished stock. The question of whether these yields are sustainable will be considered later. They comprise the "fishing mortality," but there is also a component of "natural mortality" which can be calculated indirectly from the data for some of the main species. For the demersal species this effect is quite small, but has been assumed to be about 20 percent of the total mortality (ICES, 1969). For pelagic fish such as herring, the values are higher and range from 30 to 70 percent of the total mortality. A value of 50 percent is used here (see Gulland, 1970). This gives a "total yield" of demersal fish of about 1.3×10^6, and 4.0×10^6 tons for pelagic fish.

The fishable area within the North Sea, as defined in Fig. 2.4, is about 0.5×10^6 km^2, so the total yields are 8.0 and 2.6 grams wet weight per square meter for pelagic and demersal species respectively. Given the inaccuracies inherent in such calculations, it is sufficient to convert these to calories using the factor one gram of wet weight equivalent to one kilocalorie (Winberg, 1956).

Turning to the yearly primary production, data from C^{14} measurements of production of particulate organic carbon by phytoplankton range from 70 g C/m^2 in the middle of the northern North Sea to 90 g C/m^2 for inshore areas (Steemann Nielsen, 1958; Steele and Baird, 1961 and unpublished data). These estimates do not include

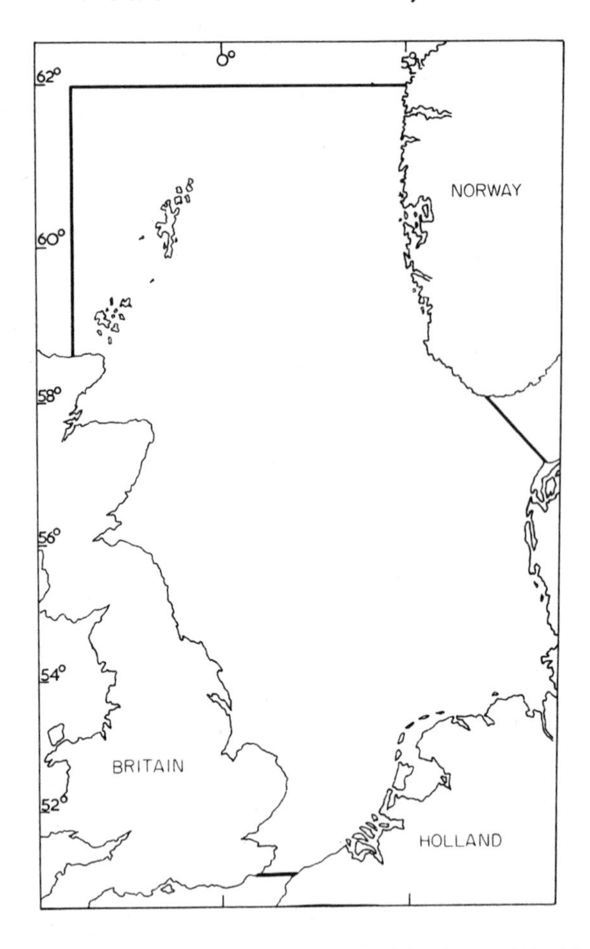

FIGURE 2.4 The area of the North Sea used for data on fish yields.

the production of soluble organic carbon by the phytoplankton. The experimental work on this extracellular production has been reviewed by Parsons and Seki (1970) and suggests that a maximum of 30 percent and an average of 15 percent of total carbon fixed by phytoplankton is released as soluble material. If this reenters the particulate fraction as bacteria with a 30 percent conversion efficiency (Parsons and Seki, 1970), then this requires a 5 to 13 percent increase in the estimate of particulate carbon production, or about 10 g C/m^2 in the

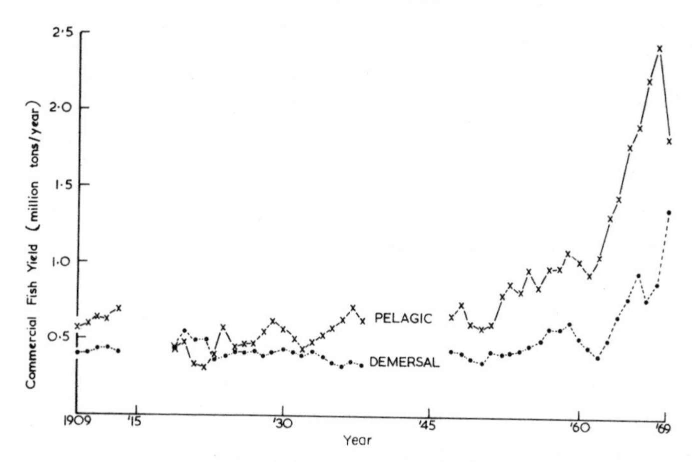

FIGURE 2.5 The annual North Sea yields of fish which feed mainly in the water (pelagic) and fish which feed on the bottom (demersal).

North Sea. The particulate organic material derived from the soluble fraction may form initially by adsorption on inorganic particles, but bacteria appear to be essential to this process (see Parsons and Seki, 1970).

Direct measurements by Duursma (1961) of soluble organic carbon in a coastal area of the southern North Sea showed an increase of 20 g C/m^2 after the spring outburst and 6 g C/m^2 in the autumn. By including estimates of rates of decomposition of this soluble material, he calculated the total yearly production as 52 g C/m^2, which corresponds to the highest observed rates of extracellular excretion. Recently, for a position in the English Channel which is not dissimilar to the North Sea, Andrews and Williams (1971) estimated the yearly intake of amino acid carbon by bacteria as 60 g C/m^2. This would be equivalent to at least 100 g C/m^2 of total carbon. They also got very high efficiencies of fixation of carbon in bacterial biomass: 78 percent for amino acids and 67 percent for glucose. Thus about 70 g C/m^2 would be produced yearly in bacteria, approximately the same as the particulate carbon production by the phytoplankton in the North Sea.

The problem with both sets of results is that the source of the soluble organic material could come not only from extracellular excretions of the phytoplankton but also from zooplankton excretions. If the source was solely extracellular plant production, then the high bacterial fixation found by Andrews and Williams would apparently "solve" the food chain problems by doubling the particulate production. However, on the basis of the experimental results already mentioned, it is difficult to accept an extracellular carbon and organic nitrogen plant production considerably greater than the particulate. Further, it is questionable whether these bacteria, which are generally less than one micron in diameter, would be directly assimilable by zooplankton such as *Calanus*. As an intermediate step, they may be eaten by protozoa (although, according to Andrews and Williams, very few suitable organisms have been found), but this would decrease considerably the energy going to the zooplankton. On the other hand, if these soluble products arise in large part from excretions by zooplankton, they do not add to the total input of organic matter to the system.

The conversion of amino acids to inorganic nitrogen compounds could be important for the recycling of nutrients to the phytoplankton, but would imply a corresponding requirement for organic carbon. It has been shown for *Calanus* during peak feeding in the spring that of the nitrogen digested, 35.7 percent was excreted in soluble form, 37.5 percent was lost as fecal pellets, and 26.8 percent was invested in growth. Of that excreted in soluble form, 88 percent was ammonia (Butler, Corner, and Marshall, 1970). Thus only 4 percent of the ingested nitrogen appears in soluble organic form unless the fecal material is broken down extensively before reaching the bottom. The latter possibility would in turn decrease the input to the benthos with a consequent aggravation of the problems in this area.

At present it is difficult to reconcile the Andrews and Williams results with other data on the nitrogen and carbon cycling in the water column. There may be a link through bacteria and protozoa to the copepods, but even with their high values of bacterial synthesis this would probably increase the food for the zooplankton by only 10 to 20 percent. Thus the main link to the pelagic fish and to the bottom fauna still appears to be through the ingestion of phytoplankton and excretion of fecal material by zooplankton. A value of 90 g C/m^2 year will be used here as an estimate of the particulate organic matter available to the metazoan herbivores. Given the uncertainties in this estimate, a conversion factor of one gram of carbon ≡ ten kilocalories is sufficient. So production is 900 kcal/m^2 year.

On this basis there are values for the two ends of the food web. Making the assumptions, which will be discussed later, that the zooplankton eat all the phytoplankton produced and excrete about 30 percent of their food as fecal material which falls to the bottom (Steele and Baird, 1972), we have the beginnings of a picture of the food web shown in Fig. 2.6. If we assume for the moment that all the fish, like the Pacific sardine, have a food requirement in calories at least twelve times their yield, then the calories assimilated by the zooplankton and those available to the benthos are roughly ten to fifteen times the food requirement of the two groups of fish. In other words, the system would fit nicely into the classical pattern if the chains were as simple as those suggested by Fig. 2.6. I have already indicated that this is not so, but I now want to go into more detail, using what we know about some of the intermediate stages in this web, to show the probable effects of increasing complexity. Also, this more detailed discussion will display the lacunae in the present stage of our knowledge of marine food webs. A general outline of the food web is given in Fig. 2.7a; each component will be considered in detail.

The herbivorous zooplankton are predominantly copepods. *Calanus finmarchicus* is an important member of this group, particularly in the northern North Sea, with the adult reaching a dry weight of about 200 μg. In coastal waters other smaller species such as *Temora*, *Pseudocalanus*, and *Acartia* can be a dominant part of the biomass. An indication of the seasonal cycles is given by the averages of numbers for the northern and southern North Sea (Fig. 2.8). These data cannot be used directly to derive values of herbivore production. All they do is to indicate that food for primary carnivores is available from April to October. Some very indirect estimates can be derived from plankton net hauls which sample the whole water column. In the open waters of the northern North Sea for which such zooplankton dry-weight data are available over a ten-year period (Adams, personal communication), the range of values between April and October is generally 3

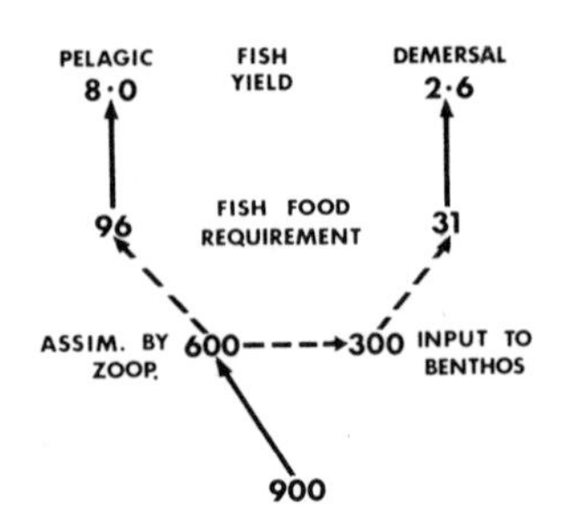

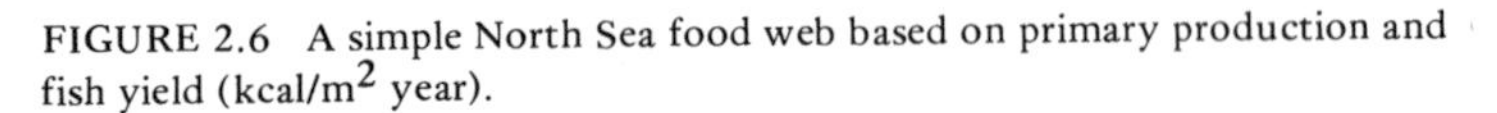
FIGURE 2.6 A simple North Sea food web based on primary production and fish yield (kcal/m^2 year).

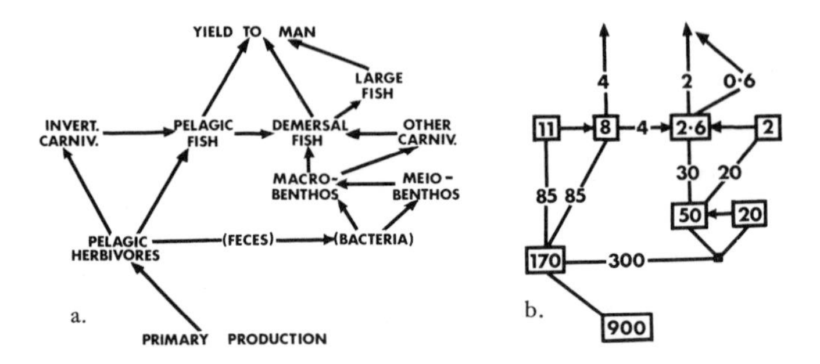

FIGURE 2.7 a. A North Sea food web based on the main groups of organisms. b. Values for yearly production (kcal/m^2 year).

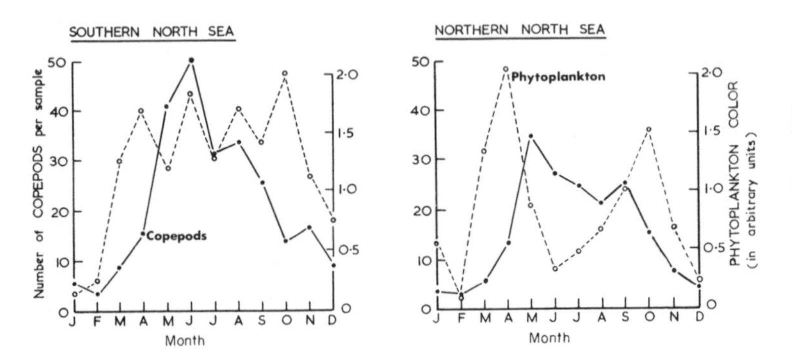

FIGURE 2.8 The long-term average of plankton cycles in two parts of the North Sea (from Colebrook and Robinson, 1961). Phytoplankton density is estimated from the intensity of green color on the silk net.

to 10 g/m^2 or, approximately, 12 to 40 kcal/m^2, with an average of about 25 kcal/m^2. The main species, such as *Calanus finmarchicus*, have three generations and with any normal predation pattern on the copepods the average population weight will be one-half to one-third of the production in one generation. Thus the total production over three generations will be six to nine times the average weight. In the model in Chapter 5 the value is seven, leading to a very rough value of herbivore production of 175 kcal/m^2 year and giving a yield to higher trophic levels of 19 percent of the primary production. This calculation obviously has a very tenuous basis, but it agrees well with the previous examples of efficiency at the herbivore level.

How does this relate to predator cycles? The herring in the North Sea are autumn spawners, so their feeding is concentrated in the period April to July (Savage, 1937). They feed for, at most, two-thirds of the period of herbivore production. Further, although according to Savage herring feed mainly on herbivores, at certain times and in certain areas pelagic carnivores such as postlarval sand eels can form up to 40 percent of the food found in the stomachs (Hardy, 1924). For this reason pelagic fish may be most important as predators during the spring. In summer and autumn other predators take over, as can be seen from the cycles of abundance of two main invertebrate carnivores, the chaetognaths (Bainbridge, 1963) and the ctenophores (Fraser, 1970), and from the increase in omnivorous euphausids in the autumn. The chaetognaths are eaten by herring and in late autumn are responsible for a slight increase in food taken by herring, above the generally low level (Hardy, 1924).

It is even more difficult to assess directly predation on the copepods by these invertebrate carnivores, but Reeve (1970) summarizing the experimental and observational data on the chaetognaths says, "If my estimate of the biomass of chaetognaths being about 30% of that of the copepods in the ocean is accepted, and that this ratio is also proportional to their production capacities . . . the majority of energy converted into animal material by copepods must be distributed to higher trophic levels via chaetognaths." This cannot be true for the North Sea, but if we take an autumn maximum population of chaetognaths of about 2,000/m^2 reaching a final dry weight of 1 mg ($\equiv$ 0.4 mg carbon ≡ 4 cal), and their efficiency in maintaining their population (as distinct from production capacity) is about 15 percent, then their combined food requirement is of the order of 50 kcal/m^2.

To emphasize how indirect this figure is, I must mention that Reeve's experimental work was done on a chaetognath, *Sagitta hispida*, from the inshore waters near Miami, because it survives well in the laboratory. Here we have an index of just how circumstantial is our present evidence, but the value for this one invertebrate predator is approximately right when compared with its consumption by herring. Values for euphausids and ctenophores would be even less reliable, but may be of the same order as chaetognaths. Thus it is reasonable to suppose that about half of the herbivores are eaten by organisms other than commercially harvestable species, which is what the numbers in Fig. 2.7b for this part of the web are intended to indicate. Again, the implication is that the herbivores must be about 20 percent efficient in transferring energy from the plants to these carnivores.

The supply of organic matter to the bottom is derived almost entirely from zooplankton feces, although a small percentage of the phytoplankton may sink out, especially in the early spring when for a short time the rate of plant growth outstrips zooplankton development. Experimental determinations of the percentage of ingested food which is assimilated vary widely (Corner and Cowey, 1968) but are mainly in the range 60 to 90 percent. If we take a value of 65 percent here (as an indication of the maximum amount of fecal material reaching the bottom in shelf areas), then about 300 kcal/m^2 is the yearly supply of energy to the benthos in the North Sea.

The way in which this energy is incorporated into the metazoan fauna which could be food for demersal fish is still largely unknown. The single most important problem concerns the role of bacteria. Newell (1965), on the basis of experiments with littoral animals, has proposed that fecal material needs to be converted into bacterial protoplasm before it is a suitable food for the fauna. Also the activities of deposit feeders can in turn stimulate the rate of bacterial metabolism (Hargrave, 1970). The efficiency of such bacterial conversion is unlikely to be higher than 30 percent and for bacterial populations living in sand is less than 10 percent (McIntyre, Munro, and

Steele, 1970). If this were an essential initial stage in the transformation of zooplankton feces when they reached the bottom of the North Sea, only 100 kcal/m^2 might be available as food for the benthos.

The benthos traditionally are divided into categories on the basis of sampling methods: the animals from grab and core samples which pass through an 0.5 mm sieve are termed the meiobenthos, and those retained are classed as macrobenthos. The macrobenthos are considered to be the main source of food for many commercial species of demersal fish. A third group, the epifauna, are defined by the fact that, because of their large size or mobility, they cannot be caught quantitatively.

In the North Sea (McIntyre, 1961) the organic carbon content of the macrobenthos ranges between 0.6 and 1.6 g/m^2. It is notoriously difficult to estimate the production by these populations. The few detailed analyses that are available (Sanders, 1956) suggest that production is two to five times the average population size. Thus the energy which they can supply yearly to higher trophic levels in the North Sea is approximately 20 to 50 kcal/m^2, which is of the same order as that required by the demersal fish populations. If the bacterial step in the food chain is included, the 31 kcal/m^2 of benthos production required by the fish (Fig. 2.7a) would have to come from the 100 kcal/m^2 produced by bacteria; this seems excessively efficient. Furthermore, it ignores the other components of the benthos.

The importance of the meiobenthos has only recently become apparent (McIntyre, 1969). Although their biomass may be relatively small (0.2 g C/m^2 for the North Sea), their yearly production may be about ten times their biomass (Gerbach, 1971). Typically the meiofauna are nematodes in deeper muddy environments and very small copepods (harpacticoids) in the interstitial spaces of sandy beaches. There are two problems associated with this group: how much of the available energy do they consume, and how far is their own production utilized by the macrobenthos? McIntyre quotes values of 10 to 30 percent intake of available energy in muddy environments and mentions evidence that the meiofauna production is not extensively utilized by higher trophic levels. In experimental work with columns of sand to reproduce beach conditions (McIntyre, Munro, and Steele,

1970) it appears that within the sand there are predators, such as turbellarians, on the harpacticoid fauna, so that effectively all the energy entering this interstitial ecosystem is dissipated within the system.

The inclusion of the meiofauna in the food web raises immediate problems. Their yearly production would appear to be of the order of 2 g C/m^2 ($\equiv$ 20 kcal). If their energy requirement is in the range five to ten times the production, it would take 100 to 200 kcal/m^2 from the available input to the bottom—roughly half the estimated input, if we ignore a bacterial step in this part of the web. If bacteria are included, the system is obviously unrealistic unless very high efficiencies are given to the transformation of detritus into bacterial biomass. The realism cannot be improved by increasing the total energy input to the bottom at the expense of the pelagic part of the web, since this is merely robbing Peter to pay Paul and would create similar problems in the pelagic phase.

The next complication is that in the simple chain the demersal fish were supposed to feed on benthic organisms that got their energy directly from the rain of organic matter to the bottom, but even this is not true. The three main demersal species—haddock, whiting, and cod—all feed to some degree on other fish; while haddock do so to a relatively small extent (Jones, 1954), cod, especially the older fish, take a major part of their diet from this source (Rae, 1967).

Examples of the stomach contents of two groups of fish are given in Table 2.2. Nearly all of the fish taken, such as sand eels, herring, and the Norway pout (*Trisopterus esmarkii*), are pelagic feeders and

TABLE 2.2 Food of fish in the North Sea; percent weight of stomach contents (Jones, 1954).

Fish examined	Food in stomach				
	Fish	Crustacea	Annelids	Mollusks	Echinoderms
Haddock (31–35 cm)	11	41	11	10	27
Whiting (21–25 cm)	32	43	16	9	–

this predation can account for the natural mortality of pelagic fish. However, this quantity is only about 4 kcal/m^2 year (Fig. 2.7b). Also some of the older, larger fish, such as cod, feed on younger stages of demersal species which can account for some of the natural mortality of these species. (There are problems in converting stomach contents into proportional feeding rates, since the larger organisms such as fish will be digested more slowly; their relative weight in the stomachs accordingly may be greatly in excess of their proportion of the diet.) The principal crustacean is *Crangon allmanni*, which is classed as an omnivorous feeder. The last three categories in Table 2.2 probably fit into the category of detritus feeders.

We are left with the third group, the epifauna, of which *Crangon* is an example. These are often predators on the benthic infauna. They are not part of the commercial fish catch but are eaten by the fish, especially cod and whiting. The same problem exists here as with invertebrate pelagic carnivores. If one allows even a small fraction of this group, say 10 percent, as fish food, then their own food requirements are of the same order as that of the fish themselves and so can roughly double the production required of the detritus-feeding benthos. The value inserted in Fig. 2.7 is intended to be an indication of this problem.

In fact, all the values in Fig. 2.7 are so tentative that they must be considered as an attempt to define problems rather than to provide answers. They do indicate, again, that transfer efficiencies around 20 percent appear to be required of the pelagic herbivores and also, possibly, of the benthic infauna that feed on fecal material. The numbers could be rearranged in various ways, but this would not alter one conclusion—that the yield of commercial fish is high in terms of the food web on which it is based.

It has already been shown from data on sardines that the harvesting of younger fish will tend to increase the yield unless there is any deleterious effect on recruitment. The same is probably true for demersal fish in terms of their energy budget (Steele, 1965b). Beyond this, however, Jones and Rae have shown that as whiting and cod grow older, fish form a progressively larger proportion of their food. It has been pointed out already that a small increase in this type of feeding has a very marked effect on yield. Thus, on the one hand, we can explain where some of the energy went before stocks were heavily fished and, on the other hand, how a decrease in the average age of fish might lead to higher yields, but also to larger population fluctuations as the food web is simplified and the life-span of the fish shortened.

Tropical Food Chains

While the examples used so far have been described in calories or in carbon, one must remember that other elements are essential in the food web. Energy can be continuously degraded, but nutrients must be recycled. In high latitudes in the North Atlantic, deep water rich in nutrients is brought to the surface by mixing each winter, thereby renewing the supply of nitrogen and phosphorus, but in the Sargasso Sea where there is a permanent thermocline, the concentration of nutrients is always low. In the latter area there may be a small rate of supply to the surface layers by a very slow upward movement of the deep water (Stommel and Aarons, 1960). In the Sargasso Sea, where light energy is always adequate, the "problem" for the ecosystem is to recycle its nutrients in the upper layers fairly rapidly rather than to pass it through a food web to longer-lived or deeper-living animals.

The estimates of net primary production off Bermuda based on C^{14} estimates of carbon uptake is about 70 g C/m^2 year (Menzel and Ryther, 1960). On the other hand, the estimate of the rate of upward movement of water through the main thermocline at 1,000 m by Stommel and Aarons is 0.5 to 1.6 cm/day. Since the nitrogen concentration* at this depth is about 20 mg at/m^3, the equivalent carbon production which could be allowed to pass back down is 3 to 9 g C/m^2 year. This may explain why measurements of zooplankton respiration in the upper waters (Menzel and Ryther, 1961) effectively balance primary production. The metabolic activity should release a corresponding quantity of nutrients back to the water so that, although

*In oceanography it is customary to express the concentrations of dissolved nutrients such as nitrogen or phosphorus in terms of atoms as mg at/m^3 (or μg at/l); this convention is used throughout.

the concentrations are low, recycling between phytoplankton and zooplankton will be relatively fast. In an area of this type, transfer efficiency of energy through the zooplankton would be low but, semantically, this is the wrong kind of efficiency to consider.

At the other extreme are areas of upwelling where nutrients are brought to the surface for considerable periods of the year with correspondingly high primary production and very high fish yields because the fish, such as anchovies, are predominantly herbivorous. For upwelling close to the coast there can be a marked lag between primary production and the onset of grazing on this plant material, so that quantities of phytoplankton fall to the bottom (Cushing, 1969). However, in a detailed study of an upwelling area further offshore near Peru, Menzel (1967) observed that there was negligible fallout of phytoplankton from the upper layers of the water, presumably caused by grazing.

Comparison with Terrestrial Systems

The ability of herbivores to crop nearly all the plant production is a universal feature of the open seas and constitutes the major difference between the marine and terrestrial ecosystems. As Crisp (1964) has pointed out, this results in part from the fact that land plants require relatively inedible components such as roots, and material like cellulose and lignin for rigidity. In consequence, even for well-managed grassland, an initial production of 5,000 kcal/m^2 gives a beef production of only 30 kcal/m^2 (Macfadyen, 1964) because only one-seventh of the primary production is eaten and only one-third of that is assimilated. Thus although the ratio of beef production to assimilation is good, the ratio of beef production to primary production is very low. Generally speaking, terrestrial primary production is usually much higher than marine primary production (Westlake, 1963), but secondary production is usually very much lower. Part of the reason for this is that populations of short-lived poikilotherms such as copepods have to invest proportionally less energy in metabolism and in overwintering (Engelmann, 1966); this is minor, however, compared with the very different structure of marine and terrestrial food webs.

These differences in levels of biomass are mentioned only briefly here since, particularly for terrestrial systems, they are discussed extensively in the references already given and in most textbooks (for example, Odum, 1971; Philipson, 1966).

There is, however, a further consequence of these differences. The structure of the terrestrial ecosystems has formed the basis for hypotheses about mechanisms for the long-term stability of these systems. I have mentioned the concept of carnivore control proposed by Hairston, Smith, and Slobodkin, among others. There are also questions of the importance of behavior patterns based on territory (Wynne-Edwards, 1962), and of the evolution of an optimum reproductive rate in terms of the ability of the parents to feed the young (Lack, 1966). These problems do not seem particularly relevant for animals in the open ocean, which release their eggs to be dispersed by water movement. Still, we may simply be ignorant of the details of life in the sea.

One other concept, diversity, is widely used in the study of all types of environment and is often associated with stability, although the definition of stability and hence the relationship are usually ill defined. At a later stage I want to use a simple trophic-level model to consider phytoplankton-copepod interactions. As a prelude, the mathematical problems of diversity in relation to stability will be considered in the next chapter.

3

Problems of Diversity, Stability, and Efficiency

In terrestrial ecology there appears to be a much greater emphasis than in marine ecology on the ideas of control and stability in populations or communities. The arguments often center around the term "density dependence." Sometimes the discussion appears to be semantic rather than ecological, but inherently the concern is to find the reasons for the persistence of species or communities in the face of fluctuations in their environment—particularly the extreme changes imposed by man when he simplifies the environment by farming or other practices and decreases the diversity of species.

Hitherto this general concept of stability has received less attention in marine ecology. However, with many problems of survival in fish stocks now arising, and with questions being asked about the possible large-scale effects of pollution, the same type of argument about the stability of community structures is very relevant to the sea. The first impression one forms of any community is usually of the diversity of species present, and of the differences in numbers, with some species abundant and others scarce. Certain general trends have been accepted: tropical regions have a greater variety of fauna than do high latitudes; marine habitats usually contain a greater wealth of species than do brackish regions (Sanders, 1968).

There have been several attempts to quantify diversity, some dealing only with the numbers of species observed, others including also estimates of relative abundance (see Pielou, 1969). These formulations provide a useful summary of descriptions of communities based on the species composition observed in the field. There does not appear to be fundamental disagreement between the formulas, and the choice of method is often a function of the sampling procedure used to compare different communities. As an example of a diversity index which involves both the numbers of species and their relative abundance, there is the following formulation, H (sometimes known as the Shannon-Weaver index):

$$H = -\sum_{j} p_j \log p_j, \qquad (3.1)$$

where p_j is the proportion of species j in the community (or, in practice, in the sample). The measure of the proportion could be numbers, or biomass, or any other common unit. This particular index derives part of its popularity from its original use in information theory, which fact, as Pielou says, has "led to false analogies that produce no noticeable advance in ecological understanding." It does, however, help to make the point that diversity as a concept is of little interest unless it can be related to other concepts such as niches (MacArthur, 1957) or to the stability of ecological systems. There have been attempts to use the Shannon-Weaver index for the latter purpose (MacArthur, 1955).

Definition of Stability

The initial problem in relation to "stability" is semantic, since there are (at least) two markedly different meanings which can be given to this term. The first use is basically descriptive, the second hypothetical. In the former we are concerned with the observed degree of persistence of parameters with time. In this sense the term "stable" can be used to define the characteristics of the physical environment of a community of organisms. (Slobodkin and Sanders, 1969, have suggested that "predictable" would be a better term, which could include environments with very regular oscillations.) The fundamental relation, for which there is considerable support (see Sanders, 1969; Dunbar, 1968), is that highly diverse communities are found where stable or predictable environments have existed for long periods of time. Under these conditions the communities themselves are stable, in the sense of persisting with relatively constant structure. This particular definition of stability derives entirely from field observation and is completely neutral on the question of which way a community would respond to perturbations of the physical environment outside the normal range of variation.

Thus the second use of the term stability arises from the hypothetical question, will a community return to its original state after it has been subjected to a particular type of disturbance? Arising from this is the subsequent question, is the degree of stability (in this second sense) directly related to the diversity of the community to which the disturbance is applied? If such a relation existed, its importance would be inestimable; it would permit us to use static "descriptions" of the present state of a community to predict the dynamics of changes which could occur in the future as a result, say, of some interference by man. The difficulty is that so far no acceptable relation has been defined. It has been suggested that high diversity could promote resistance to change (MacArthur, 1955); at the same time, it has been proposed that tropical communities which are very diverse, are also highly susceptible to disturbance (Paine, 1969).

Undoubtedly, part of the confusion has arisen through the difficulty in giving a strict definition to these terms and so in permitting a certain semantic exchange between the two meanings of stability and therefore of their relation to diversity. The trouble stems from the fact that the second definition of stability is derived from the classical theories of the dynamics of physical systems. At present it is still necessary to use this theoretical background to approach the problem. The purpose here of such an approach is to try and clear away some of the confusion, since it is apparent that argument in terms of relatively simple mathematical systems is unlikely to establish new insights. The following section will show that neither diversity nor efficiency has any relation to the dynamic stability of a system. More attention, therefore, must be paid to the details of the results of ecological experiments.

Prey-Predator Equations

Mathematically we define stability in terms of the response of a steady-state system to small perturbations in the parameter. If the system returns to its original state it is stable, and the rate at which it returns can be used as an index of the degree of stability. If the system continues to diverge from the original state after the perturbations, then it is unstable. It may go to another state defined by a different set of values for the parameters, or one or more of the parameters may go to infinity. Lastly, after the perturbations the parameters may oscillate indefinitely around the original steady-state values; this can be termed neutral stability.

The last condition is found in the classical Lotka-Volterra equations for a prey S_1 and a predator S_2:

$$dS_1/dt = (a - bS_2)\,S_1 \quad \text{and} \qquad (3.2)$$

$$dS_2/dt = (cS_1 - d)\,S_2\,. \qquad (3.3)$$

For arbitrary initial values of S_1 and S_2 the solutions oscillate indefinitely (Pielou, 1969). Because they do not converge, random fluctuations can increase the amplitude of the oscillations. Bartlett (1957), using a Monte Carlo method, has shown how random stochastic variations produce extinction in one or the other species. In this very simple relation of prey to predator, no account is taken of the time lags which normally occur in the response of one population to changes in the other. Wangersky and Cunningham (1957) have shown that they also tend to reduce the degree of stability of a system. The neutral stability of these simple mathematical relations implies that any natural system of two species operating in this way would be unstable in practice. Thus in any more complex theoretical system mathematically neutral stability would correspond to a naturally unstable system.

Of course there are, effectively, an infinite number of ways in which these equations can be made to give convergent solutions by the addition of extra terms; these methods fall roughly into three categories. It can be assumed that one population is self-limiting; for example, using the logistic growth curve,

$$dS_1/dt = (a - bS_2)\, S_1 - b'S_1{}^2 . \qquad (3.4)$$

The second method is to impose some external control; for example, by having a fixed rate of energy (or nutrient) input to S_1,

$$dS_1/dt = a - bS_1S_2 . \qquad (3.5)$$

These two categories have one feature in common: they do not involve the interaction of S_1 with S_2. This is the third category, where specific features of this interaction, such as threshold feeding levels, can produce the required convergence.

However, the question I want to consider is whether any stability can be obtained keeping the same type of interaction used in Eqs. 3.2 and 3.3, but increasing the number of interacting species. The ecological basis for this very simple approach is that in any situation where diversity indices are available, we usually do not have sufficient detail to postulate any specific feeding or behavioral pattern, and so the simplest hypotheses are those of the general Volterra equations. (Scudo, 1971, describes Volterra's work, which is followed here.)

Take a set of populations $S_1, S_2, \ldots, S_n$ (Fig. 3.1) such that for any population S_r

$$dS_r/dt = (A_r + \sum_{j=1}^{n} p_{rj}S_j)\, S_r \text{ where } p_{rr} = 0. \qquad (3.6)$$

If we assume that there exists a set of solutions $S_r > 0$ for the steady state, then the equations for small perturbations, s_r , are

$$s_r'/S_r = \sum p_{rj}s_j ,$$

with solutions of the form

$$s_r = \sum A_{rj} e^{\lambda_j t} ,$$

where the λ_j are the roots of the determinant

$$\begin{vmatrix} \lambda/S_1 & p_{12} & \ldots \\ p_{21} & \lambda/S_2 & \ldots \\ \ldots & \ldots & \ldots \end{vmatrix} = 0 .$$

This is of the form

$$\lambda^n + \alpha_2 \lambda^{n-2} \ldots + \alpha_n = 0 ,$$

so that $\Sigma \lambda_j = 0$ and any negative roots will require an equal sum of positive roots. Mathematically, therefore, s_r cannot converge and so, in the general case, random fluctuations will produce extinction of some component (gambler's ruin). Thus measures of diversity for any part of an ecosystem can have no relevance to the mathematically

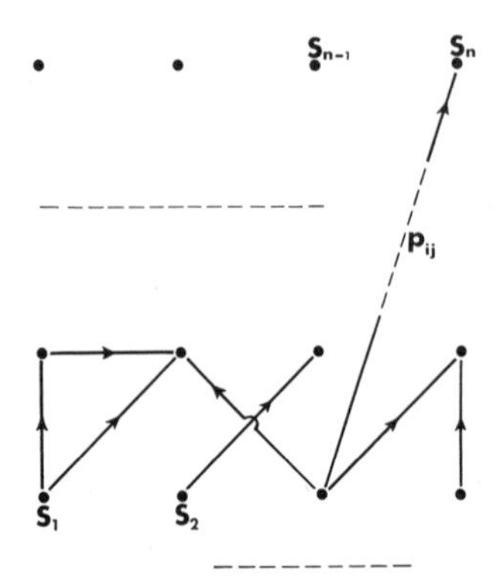

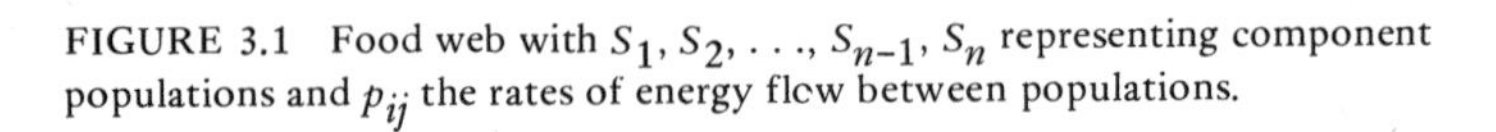

FIGURE 3.1 Food web with $S_1, S_2, \ldots, S_{n-1}, S_n$ representing component populations and p_{ij} the rates of energy flcw between populations.

defined stability of an ecosystem unless much more elaborate assumptions are made about inputs or interactions.

It is apparent that the existence of terms in the determinant of the form $(\lambda + B_r)$, where $\Sigma\ B_r > 0$, is a necessary although not a sufficient condition for convergence. The value of $\Sigma\ B_r$ could be considered as related to stability, but the measurement of this term cannot be obtained from simple ecological data. At this simple level the efficiency with which energy passes through the system does not appear to have any relation to the stability. In other words, there is no validity for the intuitive idea that energy utilized within an ecosystem by extra links in the web somehow contributes to the stability of the system (see also May, 1971).

Similarly, the horizontal links in Fig. 3.1, although they reduce the ratio of energy output at a higher level to that entering the system at the lowest level, do not affect the stability. Again this may be because of the gross oversimplifications involved, since this "ecological cannibalism" is usually time dependent in the sense that older animals of a species such as fish are more cannibalistic than young ones. The system of Fig. 3.1 can probably be stabilized in the same way as the simple prey-predator model. For example, if one assumed a basal level S_0 which could be a nutrient "reservoir" with a fixed rate of input, then the addition of an equation

$$dS_0/dt = K + \left(\sum_{j=1}^{r} p_{0j} S_j \right) S_0 , \qquad (3.7)$$

for some subset $S_1 - S_r$, should tend to stabilize the system.

Similarly, self-regulation of any one population can have this type of effect. In its extreme form, this is the Malthusian doctrine; at a more sophisticated level, the concept of territorial regulation of breeding populations limiting the numbers of young produced by that population (Wynne-Edwards, 1962) can avoid the assumption of starvation as a limiting factor, since starvation is unacceptable in nearly all ecological situations. Although the "diversity" of the web complicates the interactions and can define the levels of each component, there is no evidence to suggest that this in itself provides any extra critical features necessary for long-term survival of component species.

Spatial Heterogeneity

It has been proposed that restricted exchange between almost separate ecosystems will provide for recruitment to populations extinguished by gambler's ruin in any individual part of the system and so postpone general demise of species for an indefinitely long time. This is an extension of Huffaker's (1958) experiments, where extinction of herbivorous mite populations by predatory mites was considerably delayed by increasing the number of barriers and difficulty of access between nearly separate populations. It is supported by Simberloff and Wilson's (1969) experiments with small mangrove islands where, although a particular species might become extinct on one island, its presence on others ensured general survival and eventual return to the particular island. This "island" concept is generally difficult to accept as the sole survival mechanism and is particularly difficult to believe for pelagic populations in the open sea.

There is considerable evidence of patchiness or spatial heterogeneity in plankton populations. One can observe fluctuations in numbers of a given species on a scale of tens of meters (Wiebe, 1970) or patches on a scale of tens of kilometers which persist for several weeks (Cushing and Tungate, 1963). Examples of this type of distribution are given in Chapter 4. On the largest scales, one can define regions of an ocean within which a particular species is normally found. On the smallest scales in space and time, the fluctuations could be considered as the result of random processes of water or animal movement possibly similar to the random processes of dispersal and recolonization used as the basis for the island theory; the intermediate scales of patchiness, typified by Fig. 4.5, suggest a response linked to food availability. We do not know how such concentrations of animals on this scale, operating *against* the processes of physical diffusion, are achieved, although possible mechanisms have been proposed. For example, by combining the diurnal vertical migrations of copepods with varying current strengths at different depths and supposing that the copepods spend a longer time in the surface layers when food is plentiful, it is possible to see a way in which aggregations of copepods might be produced.

Thus patterns of feeding behavior which, in simple models, we consider as a process in time only, would in reality appear as spatial distributions. This will be discussed in more detail when the data of Fig. 4.5 are compared with the predictions of a simple model. In any general theory both random and behavioral processes leading to spatial heterogeneity would need to be included. The random processes of dispersion may have an ameliorating effect in relation to population survival (see Chapter 7) but do not seem adequate theoretically for the prevention of gambler's ruin type of extinction. Nor do they explain the patterns observed in patchiness on the intermediate scale.

Threshold Responses

On this basis neither diversity nor spatial heterogeneity in themselves appear sufficient as concepts to explain the survival of populations in the open sea. The alternative is that assumptions based on more complex interactions between individual species are necessary for any theoretical development. There are many types of possible interaction, but again the predator-prey relation is taken as the interaction where the introduction of greater complexity is likely to be most effective in producing mathematically convergent solutions. Evidence for such complexities is likely to arise from more detailed study of the behavior of individuals in response to changes in their food concentration.

A first step (Ivlev, 1961, and Holling, 1965) is to realize that an individual organism will have an upper limit to the rate at which it can assimilate food. Holling proposed the form $bS_1/(c + S_1)$ for this relation, so that the "prey" equation would now be

$$dS_1/dt = aS_1 + \frac{bS_1}{c + S_1} S_2 \, . \qquad (3.8)$$

Unfortunately, this alteration changes the pair of equations from a condition of neutral stability to one of positive divergence. Holling termed the relation, based on experiments with technicians and insects, the "invertebrate response" (Fig. 3.2a).

In other experiments mice were given a choice of varying concentrations of a preferred insect food simultaneously with an abundance of a less palatable item of diet. In this case he found that there was a threshold concentration of the insect pupae below which feeding was negligible and above which feeding increased rapidly and then leveled off (Fig. 3.2b). In a simplified form the type of relation could be expressed as $b\,(S_1 - S_0)/(c + S_1)$, where S_0 is the threshold. If this is incorporated into a pair of prey-predator equations, then

$$dS_1/dt = aS_1 - \frac{b\,(S_1 - S_0)}{c + S_1} \cdot S_2 \quad \text{and} \qquad (3.9)$$

$$dS_2/dt = \left[\frac{d\,(S_1 - S_0)}{c + S_1} - e\right] S_2 \,. \qquad (3.10)$$

Graphically, as in Fig. 3.3, the first equation represents two conditions, a lower value of S_1 where the solution is stable and an upper one which is unstable. Holling has termed this the "vertebrate response." Its importance is that under certain conditions the populations are stable. The implication is that whereas an invertebrate population would be unstable, vertebrates—as predators—could make the combined system stable. There is support for this from work by Tinbergen (1960) on birds preying on insects where the same response was found with varying ratios of one prey in relation to others.

There is an ambiguity in Holling's work, since the results can be considered to depend on experimental design as much as on the experimental animals used. Essentially his first experiments gave the animals no choice of "food," whereas the second experiments did. This is extremely important when one comes to consider herbivorous zooplankton; such invertebrates should be basically unstable in Holling's sense and he quoted work by Reeve (1963) on *Artemia* in support of his general conclusion about invertebrates. Reeve's experiments, however, used a single food source *Phaeodactylum*.

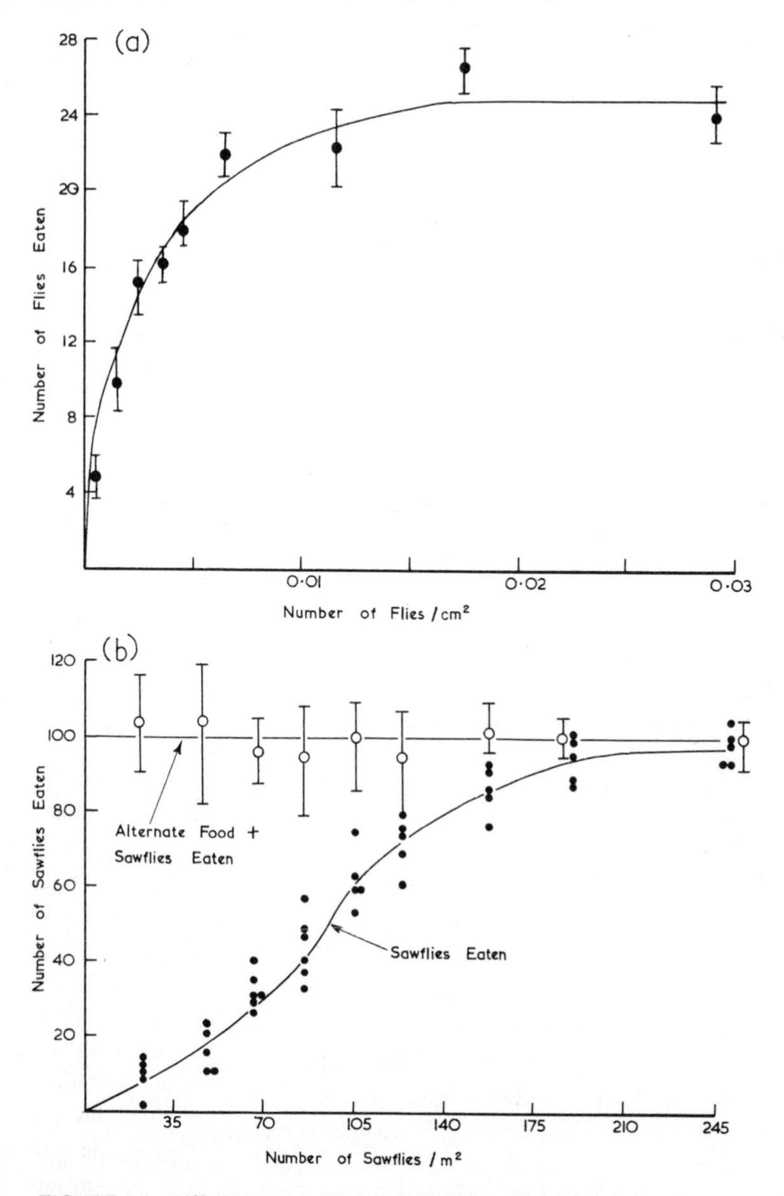

FIGURE 3.2 Differing types of functional response of predators to prey density. (a) Mantids preying on adult female houseflies; (b) deermice preying on sawfly larvae with broken biscuits as alternative food (Holling, 1965).

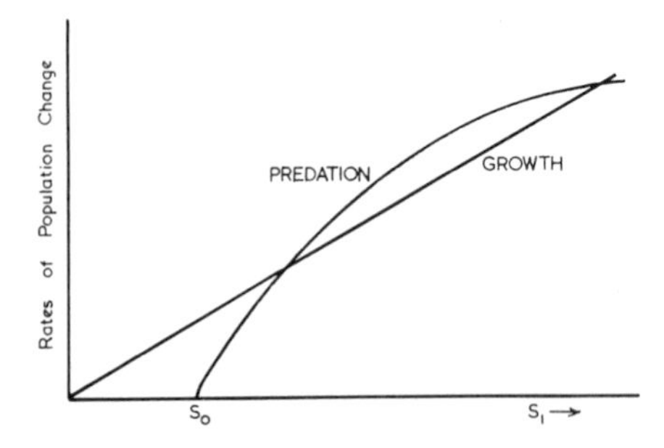

FIGURE 3.3 Schematic illustration of two possible equilibrium positions where the growth of S_1 is equal to predation. S_0 is the threshold where predation ceases.

Recently, elegant experiments by Richman and Rogers (1969) have shown that copepods can exhibit a vertebrate response. *Calanus* were fed on a culture of *Ditylum* which had the property that division of cells occurred almost synchronously. Just before final separation of most of the dividing cells the plant population contained a high proportion of food items approximately twice the size of the single cells normally present. Feeding rates on both types of cells could then be calculated for a range of proportions of each type. The results (Fig. 3.4) show a low feeding rate on the single cells, effectively independent of their percentage in the food, but a response to the larger cells which is exactly of the type proposed by Holling for vertebrates.

It is significant that the feeding rate on the paired cells was substantially higher than on the single cells, implying that there was true selection rather than merely an increased filtering rate. Harvey (1937) had already shown experimentally that *Calanus* would feed selectively at a much higher rate on a large species of diatom than on a small one, and Mullin (1963) found the same thing. Richman and Rogers' results indicate that this effect is not entirely dependent on species differences but is probably mainly a function of size.

At this point it is tempting to use these simple equations for analogies with the real world. In particular, the condition for Eqs. 3.9 and 3.10 to give a stable solution is that the coefficient e should be small in relation to d. If S_1 is a herbivore and S_2 a carnivore, this is equivalent to saying that predation on the carnivore must be slight. If it were increased markedly, the herbivore population would be unstable and would tend to become infinitely large. The analogies with examples of unsuccessful conservation are obvious and support the hypothesis of Hairston, Smith, and Slobodkin that a density-dependent relation between carnivore and herbivore may be critical for terrestrial systems.

In the sea, the experiments of Richman and Rogers would imply that density-dependent feeding can occur at the plant-herbivore level. With copepods taken from the sea and feeding on natural phytoplankton populations at different concentrations, the same type of relation has been found (Adams and Steele, 1966; Parsons et al., 1969). Parsons and his associates showed (Fig. 3.5) a complete cessation of feeding at a nonzero concentration when naturally rich food supplies were successively diluted. As they suggest, this may arise through the inhibition of selective feeding behavior at low concentrations as described in the laboratory experiments previously mentioned. The overall effect, however, for copepods feeding on a mixed food supply, is to indicate that a threshold level can be defined in terms of, say, carbon content of the food rather than requiring details of the species composition. In the simulation model to be described later, there will be this apparent contradiction; the plant food will be considered as a single unit while the feeding behavior of the copepods will be based on the concept of choice of food items. The justification rests partly on these experiments with natural populations, partly on the need, initially, to keep the model as simple as possible.

Two Prey–Two Predator

Even in these experiments with natural food supplies there is the artificiality that we use only one species of herbivore at one stage in its life cycle. The problem in the sea, and more generally, concerns the interaction of several species with overlapping food requirements. To explore the possibilities, consider two prey and two predator species where responses are as indicated in Fig. 3.6. The density-dependent stable response is that used in Eqs. 3.9 and 3.10. The unstable response is the same type of relation with $S_0 = 0$. The types of

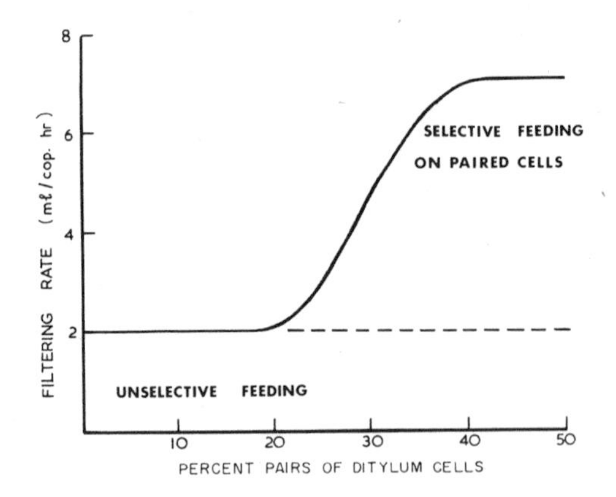

FIGURE 3.4 Filtering rates of the copepod *Calanus* at varying percentages of paired *Ditylum* cells (from Richman and Rogers, 1969).

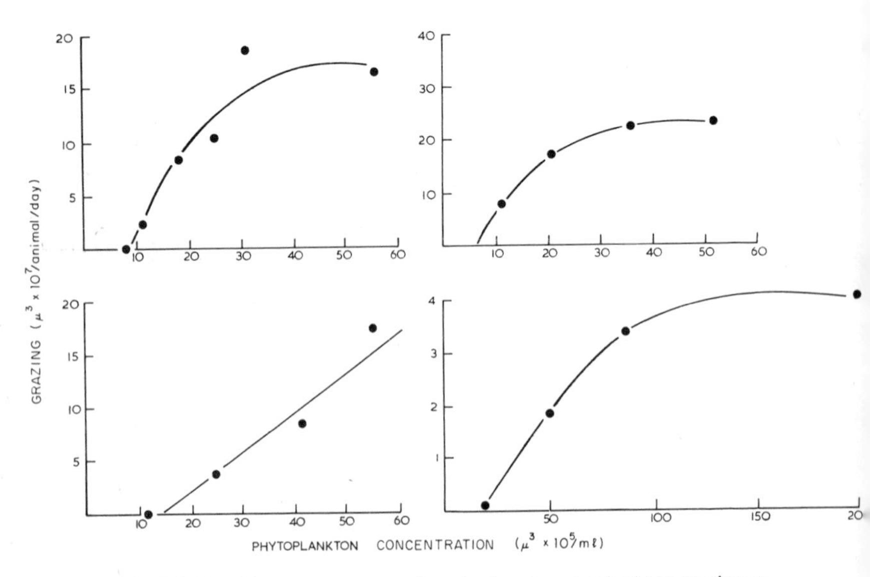

FIGURE 3.5 Four separate experiments showing zooplankton grazing at different phytoplankton concentrations obtained by serial dilution of natural populations (Parsons et al., 1969).

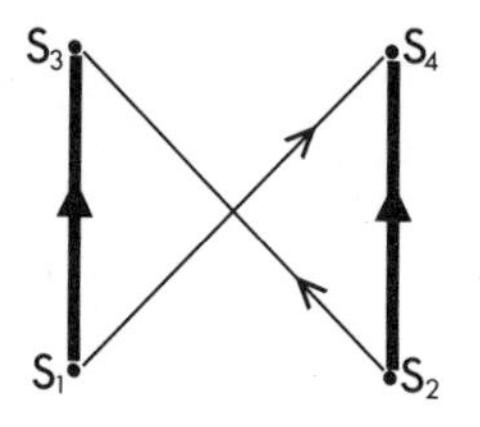

FIGURE 3.6 Interactions of two prey and two predators used for a simple model. Thick arrows denote threshold response; thin arrows, no threshold.

results obtained by varying coefficients can be shown graphically from computer runs based on the equations. The stable response, Fig. 3.7a, depends on a suitable choice of coefficients. The system can reduce to a single prey and predator, as in Fig. 3.7b, in this example by increasing the link between S_1 and S_4. Oscillations of increasing amplitude occur, Fig. 3.7c, when both the links S_1, S_4 and S_2, S_3 are increased—that is, when the energy flow through the unstable parts of the system are too large relative to the flow through the stable components. On the other hand, by removing one link completely (S_1, S_4) the system remains stable with all four components nonzero (Fig. 3.7d). A removal of either of the stable links, as one would expect, destroys the system completely. The examples illustrate the general point that a combination of different types of functional response can generate both stable and unstable systems, depending on the relative strengths of the various responses.

These results, although extremely formalized, suggest that in a complex food web it is not necessary that all feeding patterns are of a stable type as long as some elements of the predation (or grazing) conform to the general shape of response proposed by Holling, Tinbergen, and others.* This has an important corollary. It is often pointed out that even where a threshold type of feeding response exists, this predation is only a part of the cause of mortality in the prey. Thus Gibb (1966) showed that predation by tits on moth larvae in pine cones is density-dependent with a threshold in terms of num-

*In experiments such as Holling's with mice or those of Richman and Rogers, the manipulations of the experimenter in a sense take the place of the second predator in the model.

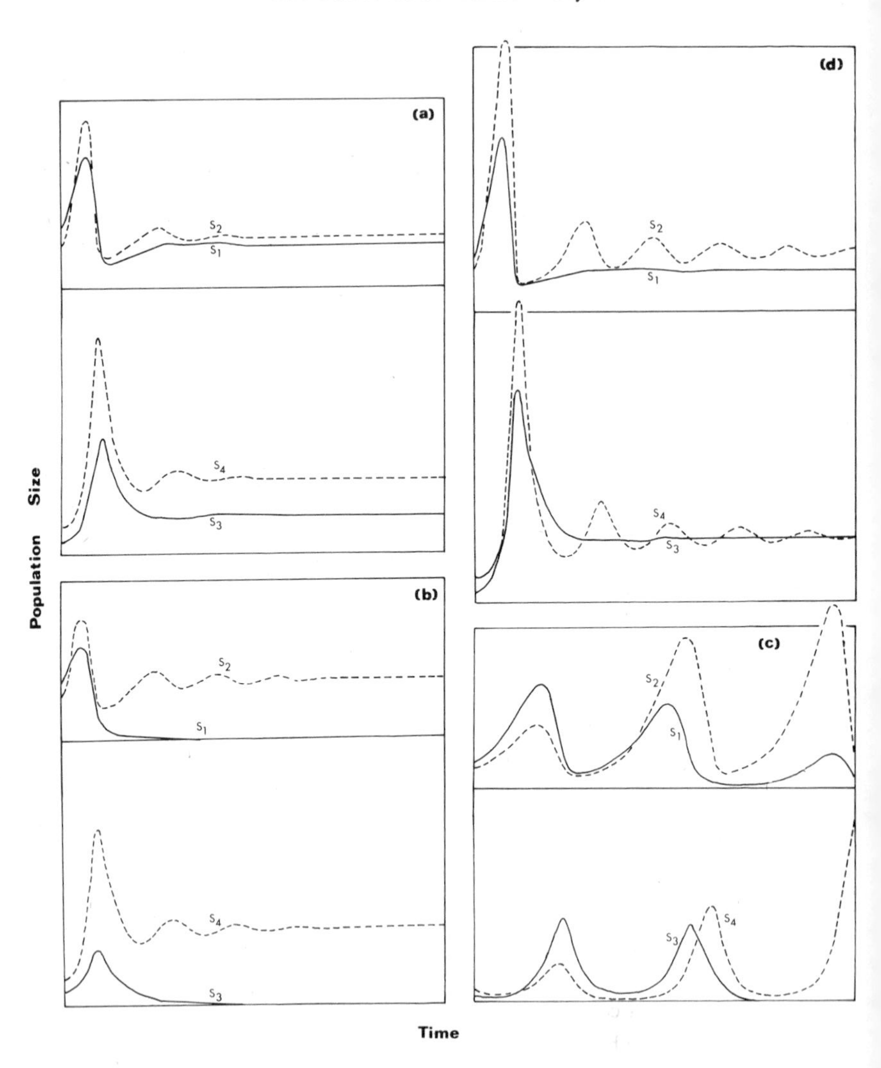

FIGURE 3.7 Computer simulations to illustrate varying responses of the prey-predator structure in Fig. 3.6 (units are arbitrary). (a) The stable response; (b) the elimination of one prey-predator pair; (c) the unstable response; and (d) the effect of removing one link.

ber of larvae per cone. He pointed out, however, that there is also a significant death of larvae when squirrels and woodpeckers destroy the cones to get the seed—a process which will be independent of the density of larvae in the cones. This very simple model would imply that, in principle, such a combination of mortality could still be effective in controlling the larvae.

Switching Responses

It may be useful to illustrate the restrictions on the ways in which these threshold concepts can be combined in food webs, since there are certain attractive combinations which, at least on this formal mathematical basis, must be excluded. Consider a predator, S_3, which in the presence of either of two prey, S_1 or S_2, would feed on them in an unstable manner (Eq. 3.8). If, however, both are present, then it will not feed on S_1 if S_1 is scarce relative to S_2 and vice versa. For a certain intermediate range of S_1/S_2 it will feed on both in a manner related to this ratio, thereby introducing a threshold type of response of the predator to each prey.

A simple mathematical representation of this is to assume that, for S_3 preying on S_1, there is an extra factor, F_1, in the predation term:

$$F_1 = \log S_1/S_2 + 0.5 \quad \text{for} \quad -0.5 \leqslant \log S_1/S_2 \leqslant +0.5\ ,$$

$$F_1 = 0 \quad \text{for} \quad \log S_1/S_2 < -0.5\ , \text{ and}$$

$$F_1 = 1 \quad \text{for} \quad \log S_1/S_2 > +0.5\ .$$

There is a similar factor for S_3 feeding on S_2. The full equations are

$$dS_1/dt = r_1 S_1 - \frac{a_1 S_1}{b_1 + S_1} (\log S_1/S_2 + 0.5) S_3\ ,$$

$$dS_2/dt = r_2 S_2 - \frac{a_2 S_2}{b_2 + S_2} (\log S_2/S_1 + 0.5) S_3\ ,$$

$$dS_3/dt = \left\{ c \left[\frac{a_1 S_1}{b_1 + S_1} (\log S_1/S_2 + 0.5) + \frac{a_2 S_2}{b_2 + S_2} (\log S_2/S_1 + 0.5) \right] - d \right\} S_3 ,$$

where the other symbols represent constants in the system. It can be shown mathematically that this is unstable, but again computer runs illustrate it best. When the system is started from arbitrary values, the two prey species converge until they reach the point where the predator has the same response to both. (In the examples in Fig. 3.8, for convenience this is taken to occur when the two prey densities are the same.) The prey then are functionally identical and the system is essentially unstable and equivalent to one prey–one predator. This possible type of system was investigated experimentally by Murdoch (1969) in the belief that it should be stable. As might be expected, no evidence of its stability was obtained, and Murdoch concluded that "switching" in nature probably is rare.

Although it may be based on ignorance of detail, in the sea the concept of functional responses as a means of control appears more acceptable than any effect arising from spatial heterogeneity. In terms of phytoplankton and zooplankton the nearest we can come to a behavioral explanation of this response is in terms of the relative sizes of the plants and the herbivores. In the North Sea there are large (*Calanus*) and small (*Temora*, *Acartia*, *Pseudocalanus*) copepods whose relative proportions can vary in an inshore-offshore gradient, or over periods of years (Steele, 1965b). The coexistence and the varying proportions could depend on the relative abundance of large phytoplankton cells as food for the larger adult *Calanus*. It could also be related to the actual phytoplankton species (Harvey, 1937; Mullin, 1963). McQueen (1970) has shown for a freshwater copepod, *Diaptomus*, that although feeding rates were low over a wide size range of unialgal cultures, high feeding rates were found for a particular size fraction of the natural food. Thus the idea of selection with a threshold level, although simple mathematically, is likely to be a consequence of a wide range of actual behavioral responses.

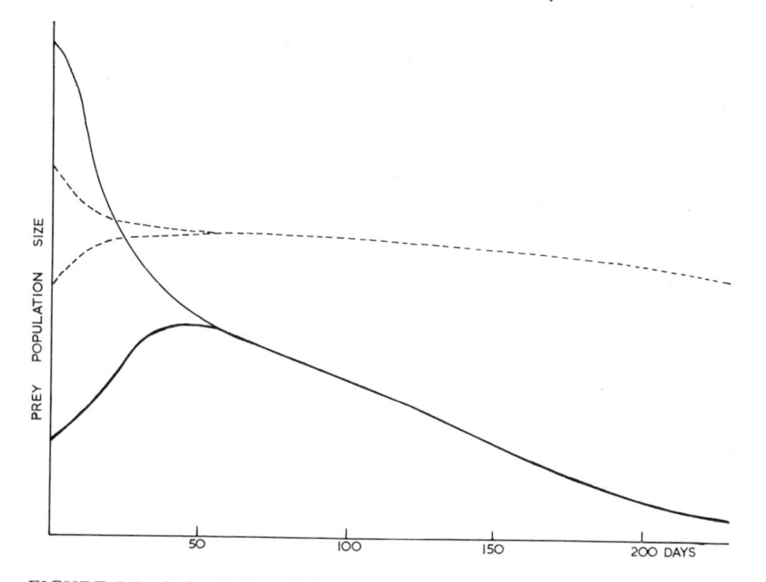

FIGURE 3.8 A simulation of the response of two prey species to one predator.

This analytical investigation of the structure of a food web is intended to indicate the type of problems that underlie any attempt to produce a detailed picture of reality by computer simulation. It explains why I prefer to use a simple food chain with complex interactions between trophic levels rather than a web with simple patterns of energy exchange.

Further, these examples show the advantages as well as the shortcomings of the analytical approach. The conclusions reached about the stability or instability of the web, the two-prey–two-predator scheme, or the ratio response are independent of the choice of parameters. The computer runs in Figs. 3.7 and 3.8 are merely illustrations. When a complex simulation model is set up, the only information one derives is from such computer runs. One can vary the numerical value of the parameters to see what happens and compare the results with observations (this can be dignified by the title of "sensitivity analysis"), but the analytical generality is no longer available. Thus, in part, the interpretation of the results of the simulation model will depend on the general conclusions reached by the simpler analytical studies.

4

Some Observations in the North Sea

The previous chapter considered some very general, perhaps excessively general, theoretical problems. At the other extreme are the data against which any theory must be tested. Much of the available information describes distributions over relatively large areas, and these can be used to point out major differences in seasonal cycles or in levels of biomass between, say, the tropics and the Arctic, or between the Atlantic and Pacific oceans. Difficulties arise when one wants to provide some explanation of the varying distributions in smaller regions of the sea over relatively short periods of time. The variability in plankton data has already been mentioned, and usually it is only by averaging over several years that any "smooth" picture is obtained. I have already given one example of such smoothed data in Fig. 2.8, which displays the types of cycle observed in the North Sea.

Another example, Fig. 4.1, shows the distribution in July of dry weight of zooplankton in the northern North Sea, averaged over a ten-year period. These data were obtained by hauls of plankton nets from near-bottom to surface, taken on research vessel cruises in the area. They can be taken to represent "summer" levels of zooplankton. The rather large intervals used in contouring are an indication of the variability that remains even with ten years of data; still, such long-term averages provide, at the least, an index of the order of magnitude of the pelagic herbivorous population. Any theory must attempt to give values within this range, but obviously this is not as restrictive a condition as one would wish. It would be possible to compare these results with data from some very different region such as the Sargasso

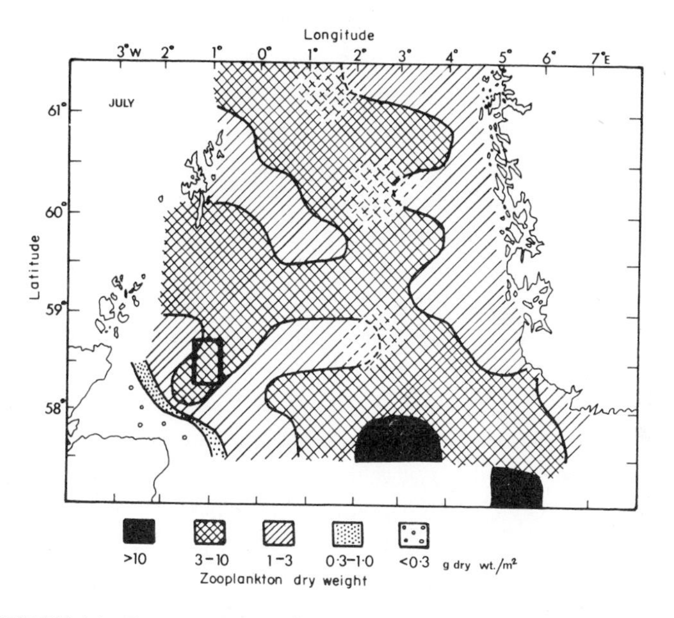

FIGURE 4.1 Ten-year average of zooplankton dry weight in the northern North Sea during July (g/m^2). The dashed lines indicate areas with inadequate data. The box shows the small area sampled in detail in July 1960.

Sea, which has a completely different cycle of production. While such comparisons are useful for "box" models of the type given in Chapter 2, they do not elucidate the problems of control I want to discuss here. For this purpose we need data on a smaller rather than a larger scale.

Results from one small area of the North Sea (Fig. 4.1) may provide some illustration of events on the scale required and also of the various difficulties in the collection of these kinds of data. The area, roughly 40 × 60 km, was surveyed twice within four days in July 1960. The research vessel steamed continuously at 6 knots and the plankton was sampled by slowly lowering and hauling a plankton net so that an oblique haul was made through the water column from near-bottom, at approximately 100 m, to surface. These hauls were started every half-hour and the time taken to complete a haul was 15 to 20 minutes so that the net was sampling the water for at least half the time.

At each half-hour, water samples were collected at 2 m depth to determine salinity, nutrients, and chlorophyll. The chlorophyll provides an indication of the concentration of phytoplankton in the upper water. Each survey (Fig. 4.2) took about 36 hours to complete.

Some vertical profiles were sampled, and two stations from the north and south of the area (Fig. 4.3) show that, although there were variations, the general features corresponded to a two-layered system. The depth of the upper mixed layer was 30 to 50 m, and this was the zone in which primary production occurred. The variations in temperature and in nutrients in the upper layer did not show any pattern and so are not considered here as an explanation of the other distributions. The main physical variable was salinity (Fig. 4.4), demonstrating that the area was not homogeneous although the variations were relatively slight and showed the same general north-south trend in both surveys. Since there is no obvious relation of the salinity distribution to the biological parameters, its possible effects are ignored. This is an example of one type of limitation on our knowledge.

The main problem is analysis of the material in the plankton hauls. The simplest presentation is in terms of dry weight (Fig. 4.5). This, in itself, produces one conclusion: that in a small area which is relatively homogeneous in the long term (Fig. 4.1), the range of values from a detailed survey on one occasion was as great as that for the long-term averages over the whole of the northern North Sea. (One point must be made about all these data. For all the nets commonly used, the mesh size permits very small zooplankton to escape. On occasion these could contribute significantly to the total dry weight of the zooplankton—for example, when the population consists mainly of juvenile stages of copepods.) A more detailed analysis of the material showed that three groups predominated: copepods, mainly *Calanus finmarchicus*; euphausids, almost entirely *Thysanoessa inermis;* and the arrowworm *Sagitta elegans*. Regressions of numbers of each group on dry weight produced conversion factors from numbers to dry weight of each group.

The data, however, showed considerable fluctuation between successive hauls. In the interval between the two surveys, eight plankton hauls were made at the same position, and from these the 5 percent

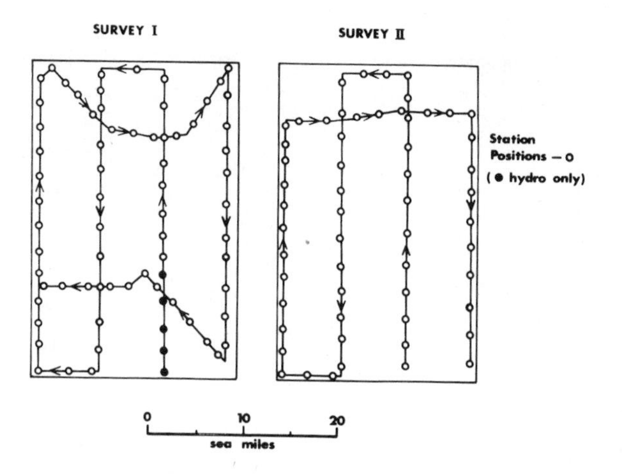

FIGURE 4.2 Track charts of two detailed plankton surveys in the North Sea in July 1960 showing sampling positions (no plankton at "hydro only" stations).

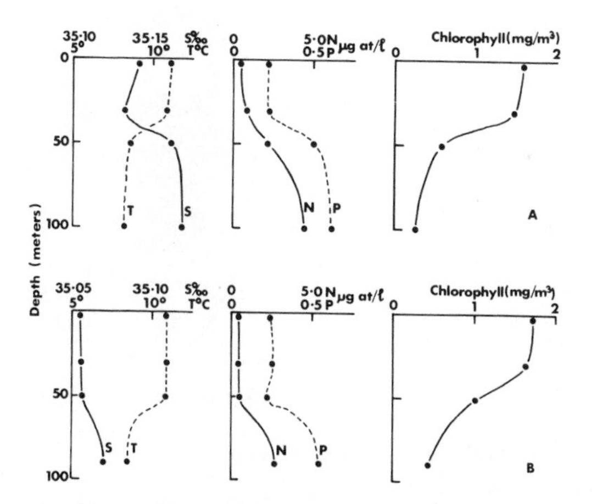

FIGURE 4.3 Vertical profiles of temperature (T), salinity (S), nitrate (N), phosphate (P), and chlorophyll a during the first survey at two stations near the north (A) and south (B) of the plankton survey area in the North Sea.

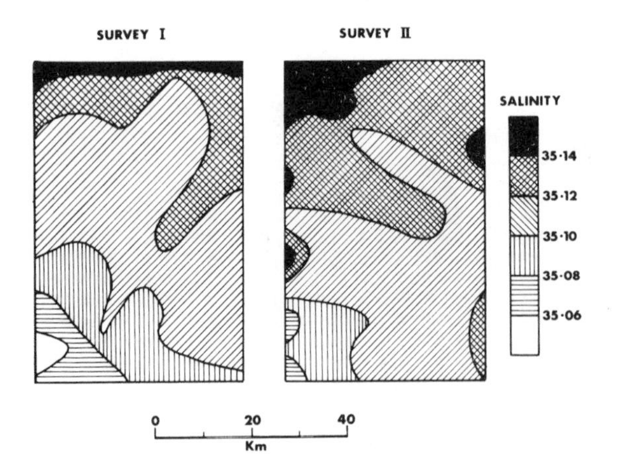

FIGURE 4.4 Distribution of near-surface salinity on the two plankton surveys in the North Sea.

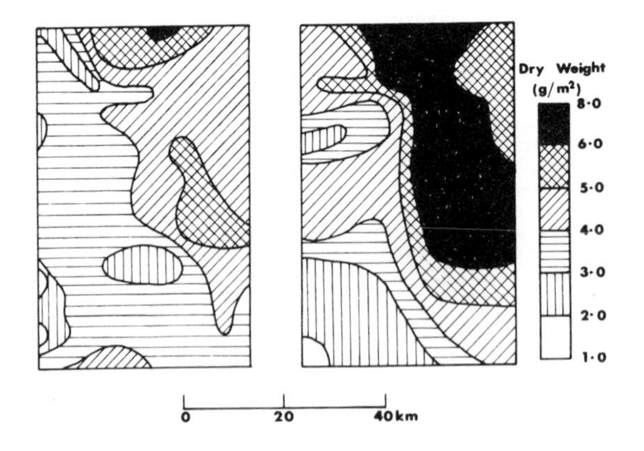

FIGURE 4.5 Distribution of dry weight of zooplankton in net hauls on the two surveys in the North Sea.

confidence limits for each value of the main components were calculated to be ±40 percent. Such a large variability accounts for the scatter in the data and makes contouring more difficult than for other variables (for example, the 95 percent confidence limits for chlorophyll were ±11 percent—Steele and Baird, 1965). It seems justifiable to take running means of the preceding, current, and succeeding

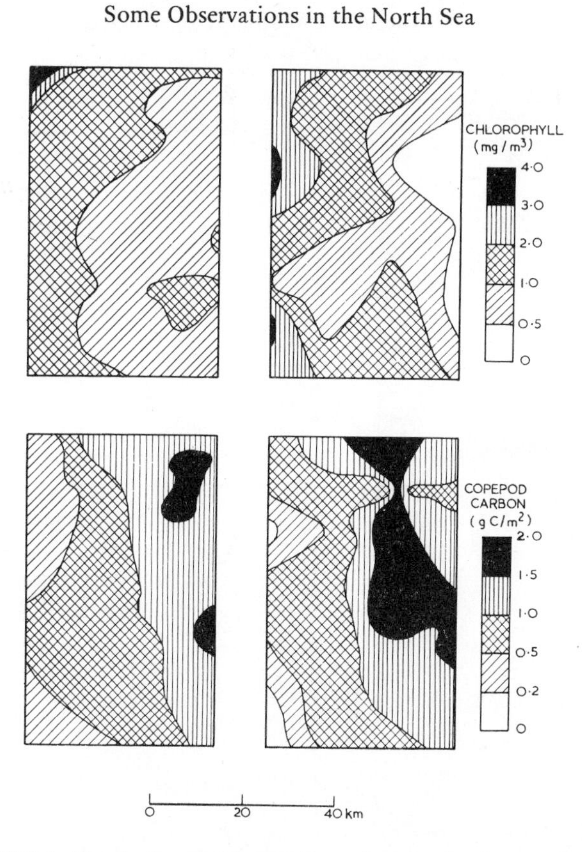

FIGURE 4.6 Distribution of chlorophyll *a* and copepod carbon on the two plankton surveys in the North Sea. (Carbon content of copepods is taken as 40 percent of dry weight.)

hauls, thereby reducing the limits to ±23 percent. Again, however, this is taking as "noise" a component of variation which ultimately may be very significant for our understanding of the details of zooplankton behavior.

The most striking feature of the charts presented as Figs. 4.6 and 4.7 is that the data show patterns definite enough that contouring was always possible and usually unambiguous. Further, each pair of chlorophyll and copepod distributions shows a significant similarity between the two surveys in terms of general east-west gradients. This would support the idea that we were observing patchiness on a scale

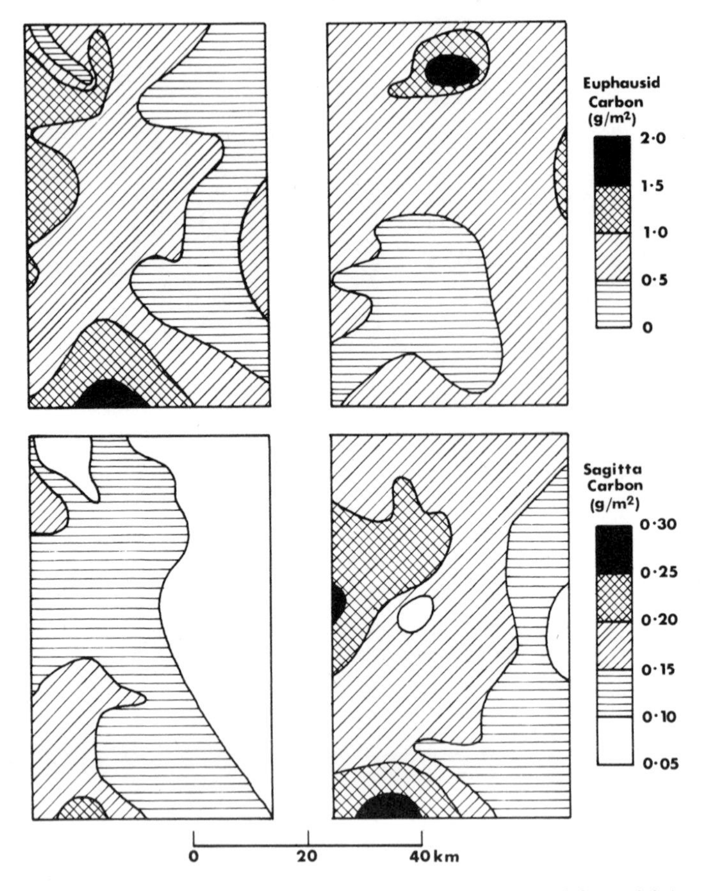

FIGURE 4.7 Distribution of euphausid carbon (40 percent of dry weight) and of remaining carbon predominantly *Sagitta* calculated by difference from the regressions. Data are from the two plankton surveys in the North Sea.

at least that of the area surveyed, and with a time scale considerably greater than the four days taken for the two surveys. There is also an impression of an inverse relation between chlorophyll and copepods, but this will be discussed in more detail in relation to the simulation model. It is apparent that a much longer time sequence is required to follow the plant-herbivore interaction. (One qualitative observation was that the copepods in the eastern sector looked much pinker, because of oil globules, suggesting that they had had a rich source of food in the past.)

With the euphausids and *Sagitta* there is no evidence of continuity between the two surveys. The change in euphausid concentration from east to west explains the increase in total dry weight to the west in the second survey, which results from this component rather than an increase in the copepod biomass. The general increase in numbers of *Sagitta* is not explicable in terms of growth over such a short period. The *Sagitta* certainly, and the euphausids possibly, were predators on the copepods. There is again a suggestion in some of the charts of an inverse relation between these species and the copepods.

The data are tantalizing because they indicate the possibility not merely of patterns in the distribution within species but of interrelations between species and their changes with time; yet even for such a short interval between surveys we do not know if water movement or diffusion could explain some of the changes. Where there is a two-layered system, as was found in this case, it is likely that water movements will be different in the upper and lower layers (Bowden, 1960). For animals which migrate vertically (and copepods, euphausids and *Sagitta* all do) the vertical differences in water velocity can provide a physical mechanism for lateral transfer of animals in or out of areas of high or low food supply. In this way temporal cycles in abundance of various trophic levels might be observed as spatial distributions in concentration during any one survey. This possibility will be used in a later chapter when the distributions described here are compared with a model based purely on time sequences.

The remaining problem, of course, is whether there is any need to "explain" such physical processes of transfer between water masses in terms of biological advantages to the population. As usual, there is a wide variety of explanations. Apart from the more obvious interpretations in terms of the consequences of grazing, Hardy and Gunther (1935) considered that copepods might be excluded from dense phytoplankton concentrations. Wynne-Edwards (1962) used the patchiness of zooplankton as further evidence for his all-embracing hypothesis that such gatherings are epideictic displays which play a part in regulating population numbers. Such widely differing explanations merely emphasize the problems of interpretation when observations are unsupported by experimental work.

There have been attempts to define theoretically the probable size of plankton patches. Kierstead and Slobodkin (1953) added a diffusion term to the simple growth equation for phytoplankton to give

$$\frac{\partial p}{\partial t} = ap + k\,\frac{\partial^2 p}{\partial x^2},$$

where x is a horizontal dimension, a is the growth rate of phytoplankton p, and k is a diffusion coefficient. They showed that if there was a small perturbation (patch) of phytoplankton of scale x_0 along the x-axis, this would increase in size only if $\pi^2 k < a x_0^2$. For a, a value of one division per day can be used as a maximum and 0.1 as a minimum. For the diffusion coefficient k, there is a problem; this is not a constant, but depends on the scale on which the turbulent diffusion is measured (Joseph and Sendner, 1958). When the scale is of the order of kilometers, k is approximately equal to x_0 (km^2/day) and the condition for an increasing perturbation is approximately $x_0 > 10/a$, corresponding to the range 10–100 km. At smaller scales, turbulence should damp out perturbations, and experimental evidence for this has been provided by Platt (1972), who showed that for scales from 10 m to 1 km the distribution of chlorophyll *a* (an index of phytoplankton abundance) had the same spectral composition as that of turbulent diffusion.

Thus patchiness arising from variable phytoplankton growth could be expected to occur in the range 10–100 km, the scale found in the surveys described here, as well as in others in the southern North Sea (Cushing and Tungate, 1963). This theory deals only with phytoplankton growth. Zooplankton grazing in patches, as shown here, must complicate the problem. When the zooplankton populations, and therefore grazing, vary inversely with phytoplankton concentration, persistent patches where concentration doubles every 5–10 km can be predicted theoretically (Steele, 1961). This compares reasonably with the patches described here.

These mechanisms can account for some amelioration of the effects of random variation in biological growth and grazing rates and could provide an indication of expected patch size. They do not explain the nature of the variations which could generate the patches. The theory and the results given here justify the need to combine large-scale observations with data collected nearer to the space and time scales usually associated with terrestrial ecology. The surveys described, and the possible interpretations, demonstrate the peculiar difficulties of sampling in the sea and the extra types of data needed: on vertical and horizontal water movements, on vertical migration of animals and its possible dependence on food, on the feeding behavior of different species, and so on.

Even if turbulence can smooth out smaller-scale perturbations, there remains the question of changes in plankton which occur on scales larger than 10–100 km. We know that, in the spring, large and rapid changes occur generally over the northern North Sea (Fig. 2.8). For these changes turbulence is an inadequate control mechanism and so responses within the ecosystem must be invoked. The simulation model presented in the next two chapters is concerned with this problem, but it must be considered in the context of the limitations in our understanding of the smaller-scale events.

5

Simulation of a Plankton Ecosystem

The simulation of a natural ecosystem is bound to involve extreme simplifications. It does not in any sense produce new facts, but merely permits the evaluation of laboratory experiments carried out on different components in isolation. By forcing one to produce formulas to define each process and put numbers to the coefficients, it reveals the lacunae in one's knowledge. Although the output of the model can be tested against existing field observations and experimental results, the main aim is to determine where the model breaks down and use it to suggest further field or experimental work.

The simulation to be developed here is part of a process to discover how necessary it is to introduce relatively complex responses to describe the interactions between predators and their prey. This model will build on the earlier discussion of threshold responses for copepods feeding on phytoplankton and use this type of response at different trophic levels. Many other aspects, however, must be simulated–growth efficiency, metabolic rate, reproduction, and so on. In this development I will be concerned not only with trying to demonstrate the necessity of a particular type of feeding response, but also with finding the limitations of such a single specific response as an explanation of the general levels of plants and animals in the sea, and so with exploring other responses. The basic defect has already been acknowledged: that the proposed interaction of a single "plant" and "herbivore" is, probably, a consequence of choices within a web. Chapter 3 suggests, however, that, if this simple system operates reasonably, extension to a more realistic web should also be possible.

As with data collection, so with models; it is possible to make general statements about energy flow in widely varying systems, or at the other extreme to examine the details of a particular system in the hope that this may be a better way to elucidate general principles. I have chosen the latter course in order that the theoretical structure may be somewhat closer to reality.

The model to be discussed here attempts to link certain features considered typical of open-sea environments in temperate or sub-arctic waters. It is probably most relevant to the northern North Sea, where the problems of sampling have been described and where an earlier, much simpler model was used (Steele, 1958). The following levels are necessary in the model:

(a) The vertical mixing of water through a stable thermocline. This is a process which lasts for about two hundred days in the northern North Sea and which transfers nutrients upward into the upper euphotic zone where plant production occurs, and also mixes some of this plant production downward.

(b) The nutrient cycle in the upper mixed layer. This cycle is a balance between supply by vertical mixing and herbivore excretion, and consumption by plants. The initial condition is a relatively high nutrient concentration after the winter period of low plant production. The biological problem is the relation of the rate of nutrient uptake and growth kinetics of the plant population to the nutrient concentration.

(c) Grazing by herbivorous copepods, considered the main factor in removing the plants produced. There are three parameters: the relation of grazing rate to food concentration; the ratio of assimilation to ingestion; and respiration rate, which may be related to feeding rate.

(d) The growth rate of the copepods, determined by these parameters plus a choice of initial weight and of a final weight of the adult. When the latter is reached, it is assumed that extra food above the metabolic requirement goes to reproduction for a given period of time at the end of which juveniles are produced and the cycle restarts.

(e) The predation rate on the herbivores, the most difficult factor. It has been decided to close what is really an open system by testing different assumptions about predation rate as a function of herbivore weight and numbers, without specifying the actual predators or considering their population dynamics.

The formal derivation of each of these parameters will now be described in detail. This can be regarded as a rather selective review of the problems which exist in quantifying the various aspects of a pelagic marine food chain.

Physical Parameters

Vertical mixing. The formation of the shallow thermocline in spring usually induces an outburst of phytoplankton (see Fig. 2.8). However, in any year this process may be irregular with a few short-lived thermoclines forming in calm periods and then breaking down before the establishment of the main thermocline, which lasts throughout the summer and autumn. These transient thermoclines will give short bursts of plant production which could trigger the change of the small population of copepods from their overwintering preadult stage into reproducing adults and so start the production of juveniles, which is the next main feature of spring conditions. Once the main thermocline is established, the depth of the mixed layer above it decreases to a minimum about July-August and then increases with cooling of the upper layer in the autumn until the whole water column becomes isothermal and plant production effectively ceases. In midsummer the nutrients are nearly exhausted in the upper layer (Fig. 5.1a).

Although these features can be simulated, they would tend to complicate the interpretation of the biological changes. Thus, as in the earlier model (Steele, 1958), a fixed thermocline depth of 40 m is used and a mixing rate, V, is defined as the fraction of the upper layer exchanged daily with the lower layer (Fig. 5.1b). Variations in this rate correspond to different parts of the North Sea. The rather sporadic events at the start of production will have their main effect on production rates of the phytoplankton. Although this could be simulated by stochastic fluctuations, for the present our test will be to see how the model responds to a range of plant production rates.

Light and temperature. The average light available in the upper mixed layer will depend on the depth of the mixed layer, as well as on day-

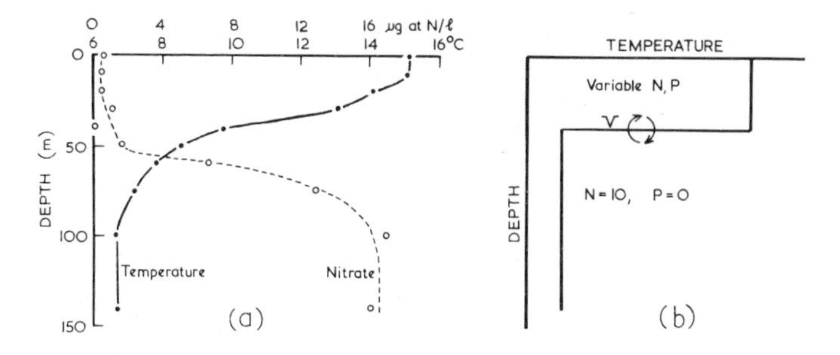

FIGURE 5.1(a) Temperature and nitrate profiles in July 1965, in the center of the northern North Sea; (b) the idealized temperature profile with some of the variables.

to-day and seasonal variations. I have suggested earlier (Steele, 1965a) that the rate of solar energy input required to form the thermocline is sufficient to ensure that light for photosynthesis is not a seriously limiting factor thereafter. The effects of day-to-day fluctuations in radiation were investigated with a simple model (Steele, 1958) and appeared to be smoothed out at herbivore level (Steele, 1961). To study this further, we need detailed data on daily values of light and thermocline depth; until these are available it is simplest to assume that light, like thermocline depth, is not a variable.

Similarly, temperature changes will certainly affect plant species composition and animal metabolism, but for an area like the North Sea where the range is only 7 to 13° C this is not likely to be important for overall stability and so is ignored.

Chemical Parameters

Nutrient uptake kinetics. Many nutrients such as phosphorus, nitrogen, and silicon, and also metabolites such as B_{12}, can limit the growth of phytoplankton. It is now generally accepted, however, that nitrogen is normally at least as limiting as any other component. Recent work therefore has been on the effects of varying concentrations of nitrate and ammonia, the usual forms in which nitrogen is available to plant cells in the sea. Short-term experiments on the rate of nitrogen uptake may not correspond exactly to growth rates of cellular material, but they indicate the type of response. In nearly all cases

(Eppley and Thomas, 1969; Eppley, Rogers, and McCarthy, 1969; MacIsaac and Dugdale, 1969) this response to a nitrogen concentration, N, in the water can be fitted by the relation

$$AN/(B + N)\,, \qquad (5.1)$$

where A represents the maximum rate and B is the nutrient concentration at which half the maximum rate is obtained. MacIsaac and Dugdale summarize their results with natural populations as follows: "A low value of B would be of competitive advantage to species found in nutrient poor waters, and especially for nitrate such a situation is found. The associated measurements of A are also low. The eutrophic stations are all characterized by higher values of A and B." These two types of response are illustrated in Fig. 5.2.

The problem here is that when the nutrient concentration changes, as in Fig. 5.3, the plant species present do not adapt to the new concentration; rather, the species composition alters and the new dominants are probably species with a different response which gives them a competitive advantage. Since, for simplicity, I wish to consider as a function of nitrogen the overall rate of production by photosynthesis, rather than the response of particular species, it is the envelope of the individual curves that is significant (indicated by the dashed line in Fig. 5.2). The shape of this curve is unknown, but the simplest assumption is to give it the same general form as the species curves. Thus the relationship given in Fig. 5.1 above will be used to define the rate of nitrogen uptake by unit cell mass. By varying A and B, we can test the sensitivity of the whole system to the exact shape; this test will also include the effects of changes in any factor such as light, thermocline depth, and the like, which alters the rate of primary production.

Nutrient mixing. It is assumed that at the start of the spring outburst there is a nutrient concentration, NO, throughout the water column. After the column is divided into two layers, mixing into the upper layer by the exchange rate V is $V(NO - N)$.

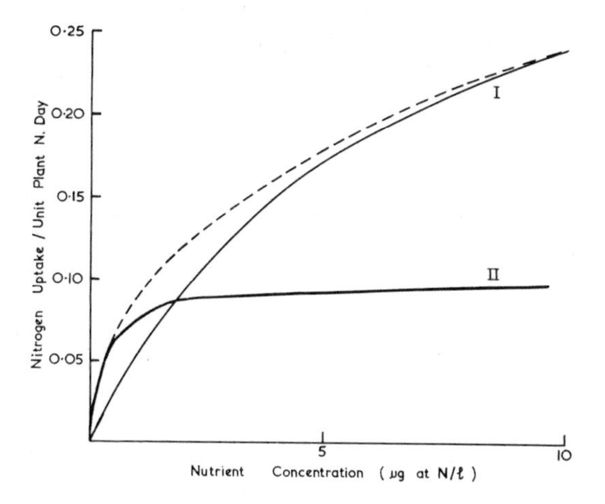

FIGURE 5.2 Examples of the differences in nutrient uptake that are found for plants taken from a high (I) and low (II) ambient nutrient environment, and the possible envelope for such responses.

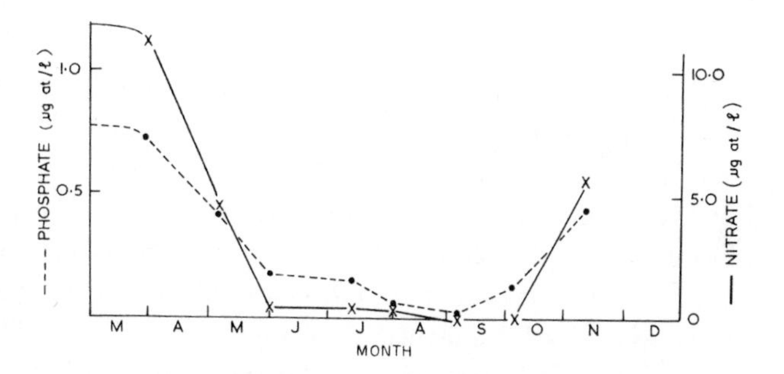

FIGURE 5.3 The cycle of nitrate and phosphate concentrations in the upper layer of the water column in the center of the northern North Sea.

Nutrient regeneration. The simplest assumption we can make about biochemical composition is to take the proportions in plants and animals as the same. Further, instead of a general term "nutrient," nitrogen is taken to be at least as limiting to plant growth as any other factor. The plant nitrogen grazed by the herbivores is assumed to be divided into assimilated nitrogen and nitrogen in fecal pellets, and the proportions are the same as the carbon content of these two fractions. The fecal pellets are lost to the upper system and supply the bottom ecosystem. Soluble nitrogen excretion is taken as being in proportion to metabolic rate, and although there has been some disagreement, it is assumed that the zooplankton are ammonotelic (Corner and Cowey, 1968) and so the excreted nitrogen potentially is immediately usable by the plants. However, part of this excretion certainly occurs below the euphotic zone and is not in practice available for phytoplankton. The fraction which is available, U, will be varied to test the importance of this factor.

Biological Parameters

Zooplankton grazing rate. Considered the most critical factor in the model, the zooplankton grazing rate has been discussed in a previous chapter. The most suitable formal expression for this factor is

$$C(P - P1)/(D + P)\,, \tag{5.2}$$

when P is the concentration of particulate carbon (μg C/1), $P1$ is the threshold concentration below which feeding is zero, and C and D are parameters defining the shape of the curve. By taking $P1 = 0$, the original concept of filter feeding is simulated and the model can be used to test the consequences of both hypotheses. The effects of varying concentrations of P on quantity of carbon ingested and on "volume swept clear" are shown in Fig. 5.4.

A further problem concerns the type of particulate carbon represented by the term P. In addition to living phytoplankton cells, seawater contains organic debris which can form 10 to 90 percent of the particulate carbon depending on the richness of the phytoplankton populations. This "detritus" has played a contentious role in recent discussions of the cycling of organic matter in the sea. It has been proposed as a major food of copepods (Riley, 1963), yet attempts to get *Calanus* to eat it have been unsuccessful (Paffenhofer and Strickland, 1970). The relative abundance of detritus is, in fact, an argument against any heavy grazing on it (Steele, 1965a).

It would be possible theoretically to include detritus as a relatively

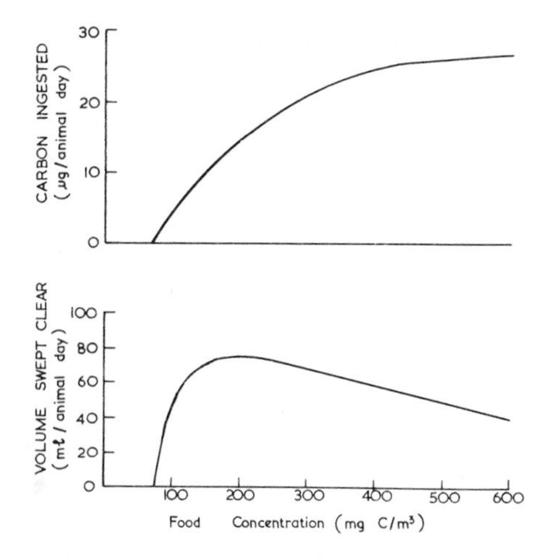

FIGURE 5.4 The theoretical relation used to describe carbon ingested as a function of carbon in the food, and the derived volume swept clear.

unsuitable source of food, rather like the broken dog biscuits in Holling's experiments with mice feeding preferentially on saw-fly larvae (see Fig. 3.2b). This source might then become important for the small populations of overwintering copepods in temperate regions, or for copepods living primarily in deep water. Such situations are not being considered at present and the detritus is regarded as outside the system, although some 50 μg C/l should be added to the calculated carbon for comparison with observed values.

Similarly, plant material sinking out or carried downward by mixing is ignored, but since plant concentrations below the euphotic zone are very low, feeding in this region must be minimal. Paffenhofer and Strickland showed that fecal material could be reutilized, even by the species that produced it. However, fecal pellets have a fast sinking rate, generally between 100 and 300 m/day (Smayda, 1969). This may exceed the vertical migration rate of the zooplankton (anyway, we want this material on the bottom as food supply for the benthos). Lastly, vertical migration of the zooplankton is ignored. The reasons for this daily movement in and out of the upper feeding layer are still uncertain, although various metabolic advantages to the animal have been proposed (McLaren, 1963; McAllister, 1970). Calculations suggest that the energy expenditure on vertical migration is negligible (Vlymen, 1970). The feeding rate used here is that applying to average food intake over twenty-four hours and the food, P, is assumed to be phytoplankton from the upper mixed layer.

Assimilation. As a fraction of food ingested, assimilation may vary with food concentration and in particular the fraction may be small when food is abundant (Beklemishev, 1962). This has not been supported by experimental work at natural food concentrations (Conover, 1966). The introduction of an assimilation/ingestion ratio inversely proportional to food concentration would tend to smooth out the response of zooplankton to food changes. The alternative assumption of a fixed ratio is likely to be less stabilizing and is adopted here by taking a fixed value of 0.7 for the ratio, rather than by adding another dubious variable to the rapidly lengthening list.

Respiration rate. The simplest assumption is that respiration rate depends only on temperature for a given weight of animal; since temperature changes are ignored here, this would imply a fixed rate, F. However, if one accepts that a copepod varies its feeding behavior in response to changes in food concentration, it would be ecologically foolish to maintain a fixed metabolic rate. Small flatfish (Edwards, Finlayson, and Steele, 1969) in laboratory experiments can increase their metabolic rate in proportion to the rate of food intake. Mullin and Brooks (1970b) have provided indirect evidence that *Calanus helgolandicus* makes the same kind of adaptation. This would imply an ability, particularly of overwintering animals, to make relatively rapid metabolic adjustments, as distinct from seasonal changes which may be related to food (Marshall and Orr, 1958; Conover, 1962). The most extreme assumption is to take a second respiration rate, E, directly proportional to food intake. The balance of gain by assimilation and loss by metabolism (where E and F are carbon equivalents) is then of the form

$$(0.7\,C - E)\,(P - P1)/(D + P) - F\,.$$

Relation of copepod weight to feeding and metabolism. It is generally assumed that the effect of weight variations on metabolism is of the form W^a, where a lies between 0.6 and 0.8. In studies of farm animals it is common practice to express food intake in terms of this "metabolic weight," W^a, (Blaxter, 1962) and it has proved appropriate also for small flatfish (Edwards, Finlayson, and Steele, 1969). The relation appears to hold for different sizes of copepods (Gauld, 1951; Mullin and Brooks, 1970a). Thus a physiologically acceptable assumption is to take both metabolism and feeding rate as proportional to $W^{0.7}$. The equation for the rate of change in weight (expressed in carbon units) is then

$$DW/DT = \left[(0.7C - E)(P - P1)/(D + P) - F \right] W^{0.7}.$$

The excretion rate (as carbon equivalent of nitrogen released) is

$$\left[E(P - P1)/(D + P) + F \right] W^{0.7}.$$

Initial and final copepod weights. At this point it is necessary to decide what "type" of copepod will be used in the model. Since *Calanus* species are often dominant in the North Sea and other regions and since nearly all experimental data refer to them, they will be used as the prototype. The initial weight is the egg, but the relevant value is the carbon content of the nauplius when it starts feeding. Data on calanoids (Mullin and Brooks, 1970a) show that Nauplius I has about 0.2 percent of the carbon content of the adult. In the North Sea, 100 μg C is a rough average carbon content for an adult (Marshall and Orr, 1955), so that the initial value is taken as 0.2 μg C. According to these authors the observed weights do vary seasonally by about ±20 percent, and Mullin and Brooks (1970b) have shown that adult weight depends on food concentration during growth, although growth efficiency is not affected. I have chosen to ignore this factor.

It is assumed that once a copepod reaches the adult stage (VI), growth ceases and food intake above metabolic requirement is used for reproduction. Although adult males have a much shorter life span than females, their food requirements should be included in the energy used for reproduction, even though the egg production by the females predominates. If the egg, initially, is mainly yolk and the efficiency of assimilation by the developing nauplius is about 30 percent, then a factor of 0.3 can be used to convert adult female food intake (in excess of metabolism) into juvenile stock. This fits reasonably with Marshall and Orr's estimate of the size of *Calanus* eggs. There is also likely to be some special mortality at the point when the nauplius starts feeding, separate from other predatory factors. While there is no evidence on this point at present, the results of the model will indicate that it is important as a factor.

On this basis, reproduction of the female population is simulated by accumulating excess food over a fixed period of time, J, corresponding to the length of the female life-span; multiplying this quantity by a factor X ($\leqslant 0.3$) to get the total carbon content of the juveniles; and, finally, dividing by the initial weight to give the number of juveniles which then begin a new cycle of growth. This process means that there is only one cohort of animals, all of the same weight. It is possible, however, to introduce successive cohorts into the model and a "run" with two cohorts will illustrate the interactions that could arise from two bursts of egg-laying at the start of the spring outburst of plant growth. Larger numbers of cohorts or a semicontinuous egg-laying might have been used; although more realistic, such a procedure would complicate the study of variations in parameters. Since the presence of only one or two cohorts is likely to be the least stable situation, the effects of variations in the parameters are emphasized.

Herbivore mortality. Any ecosystem is open-sided and open-ended, but for a simulation such as this a closed system has to be used. I have already discussed the reasons for closing the sides by using a simple chain. The lower end has been closed by ignoring the complex processes of nutrient regeneration which occur in deeper water or on the bottom. The top end, however, is the most difficult to close. There are several types of predator on the herbivores (as discussed in Chapter 2) and in principle the dynamics of these populations should be included. In order to close the system, I have ignored them and merely considered different types of predatory behavior.

The simplest type of mortality is a death rate proportional to the numbers in the population at any instant, independent of population size and the individual weights of the organisms. This is the type of predation on herbivores used in the Lotka-Volterra equations of Chapter 3 and will be defined here by a coefficient GX. The alternative assumption is that mortality tends to occur over relatively short periods in the life of an organism. One of these is the end of the yolk-sac stage. For demersal fish such as cod and haddock, which produce very large numbers of eggs, it is usually assumed that between egg production and postmetamorphic settlement on the bottom, survival is of the order of one in 100,000. Egg production by an individual female *Calanus* is probably in the range 100 to 1,000 (Marshall and Orr, 1955). This value is not much greater than the ratio of numbers of preadult copepods after the spring bloom to the overwintering number of parents before the bloom. Thus it is possible that, normally, mortality at this period is not so dramatic as with the corresponding stage in the life cycle of fish. However, the results of the model will suggest that a simple percent mortality, introduced by making $X = 0.1$, is insufficient for all possible environmental conditions.

After the early naupliar stages there is some evidence that mortality may be low until the last preadult copepodite stage (V), at least in calanoids from relatively rich areas. There have been two detailed studies of the growth of such natural populations. Cushing and Tungate (1963) sampled a *Calanus* patch in the North Sea as it developed. From the observed numbers (Fig. 5.5a) I have estimated the population size as it passes through each stage (Fig. 5.5b). This is based on the proportions of its life cycle that *Calanus* spends in each stage according to Marshall and Orr (1955). The general conclusion, that mortality is negligible until stage V, depends only on the proportions of time rather than on the absolute periods of time. Parsons and his co-workers (1969) reached similar conclusions for the development of a *Calanus* population in the coastal waters of British Columbia. In both cases it is likely that fish are the main predators of the copepods and, as suggested by Parsons, they may have a preference for the larger stages of the copepods and will search for relatively dense concentrations of them.

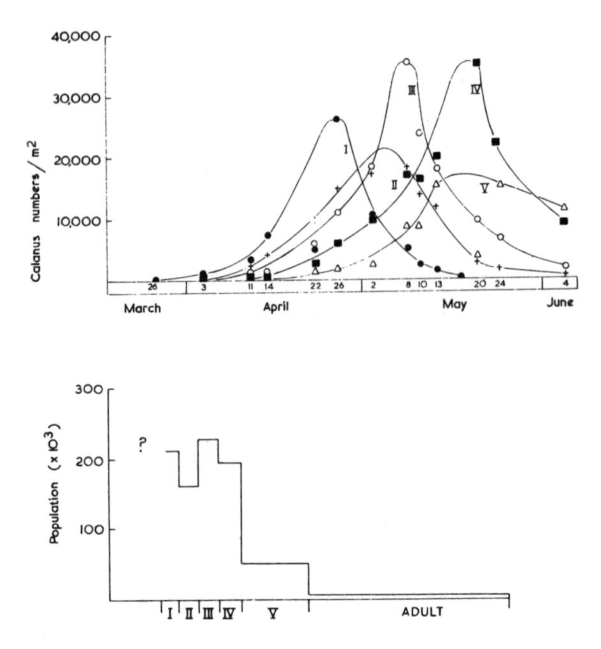

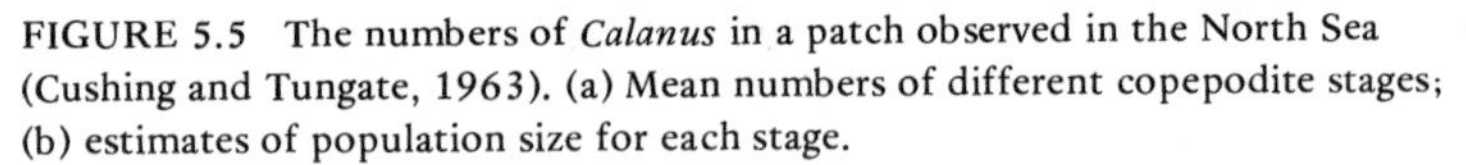

FIGURE 5.5 The numbers of *Calanus* in a patch observed in the North Sea (Cushing and Tungate, 1963). (a) Mean numbers of different copepodite stages; (b) estimates of population size for each stage.

Thus predation could depend on the combination of number, Z, multiplied by weight, W, above a certain minimum weight, $W1$. At the other extreme, the very small overwintering populations of stage V copepods appear to survive for 5 to 6 months with little mortality so that there may be an extremely low threshold of numbers, $Z1$, below which there is no predation. Cushing's (1959) value for *Calanus* of $1{,}000/m^2$ is used. This is equivalent to a population of 0.1 g C/m^2, which is approximately that observed in the North Sea in winter. Lastly, for very high concentrations of large copepods some relative decrease in proportional removal rate may be expected because of a limit on the concentration of predators or the rate at which they can handle their prey.

In summary, the predation rate on a copepod population, Z, of individual carbon content, W, is

$$DZ/DT = -G(Z - Z1)(W - W1)/(H + Z \cdot W) - GX \cdot Z\,,$$

where G and H are parameters defining the shape of the curve for the second type of predation. This relation, like that for the herbivores, involves thresholds $Z1$ and $W1$ which can be given a range of values including zero. In this way, the importance of such a predator response can be studied in a marine system.

Food chain efficiency. This term has had a variety of definitions, but is a convenient way of summarizing the cumulative transfer of energy in a system. At any date in a computer run, "efficiency" will be the ratio (using carbon units) of the total predation on the herbivores from the start of the run to the total primary production from the start of the run.

Units

Even though it would be more consistent to use energy units, nearly all the available data are in terms of carbon and nitrogen. Thus I have used carbon as the common unit for all the trophic levels. It has been proposed (Redfield, Ketchum, and Richards, 1963) that the proportion of elements in plankton is relatively constant with a mean C:N ratio of 5.4:1. For eleven species of phytoplankton Parsons, Stephens, and Strickland (1961) found ratios between 4.3:1 and 9.0:1, with an average of 6.1:1. Sargasso Sea zooplankton have a ratio of 5:1 (Beers, 1966) and *Calanus* is about 6:1 (Marshall and Orr, 1955). For this model the value of Redfield and his colleagues is used. The nutrient term, N, can then be converted to a carbon equivalent, R, and the concentration of nutrients available in the winter given as RO. In the presentation of the data, however, nitrogen as μg at/1 is still used.

Initial Conditions

The main question posed is how stable the model system would be when subjected to perturbations, and whether the levels reached after such perturbations are of the right order of magnitude when compared with field data. From this point of view the bloom of plankton in the spring is the largest type of perturbation which occurs naturally; the attempt to simulate it therefore should be a suitable test of the assumptions on which the model is based.

The mechanism that triggers the change of the overwintering stage V into adults in early spring is not exactly clear. It may be caused by the first small pulse of plant production resulting from a transient thermocline. This pulse is assumed to be sufficient to produce a number $Z0$ of juvenile copepods of weight $W0$. The model is started with this population and with a plant concentration $P0$, which is the threshold at which feeding occurs.

Equations

Our set of equations can now be written formally in terms of the daily change (D/DT) of each parameter. For one cohort:

$$DR/DT = -A \cdot R \cdot P \cdot /(B+R) + V \cdot (R0-R)$$

$$+ U \cdot \left[E(P-P1)/(D+P) + F\right] Z \cdot W^{0.7}$$

(Nutrient change = −plant uptake + mixing + zooplankton excretion);

$$DP/DT = A \cdot R \cdot P \cdot /(B+R) - V \cdot P - C \cdot Z \cdot W^{0.7} \cdot (P - P1)/(D + P)$$

(Plant growth = nutrient uptake − loss by mixing − zooplankton grazing);

$$DW/DT = \left[(0.7C-E) \cdot (P - P1)/(D+P) - F \right] \cdot W^{0.7}$$

(Zooplankton growth = assimilated food - respiration); and

$$DZ/DT = -G \cdot (Z-Z1) \cdot (W-W1)/(H+Z \cdot W) - GX \cdot Z$$

(Zooplankton numbers = density-dependent predation − proportional predation.

Terms containing $P-P1$, $Z-Z1$, and $W-W1$ become zero for $P \leqslant P1$, $Z \leqslant Z1$, and $W \leqslant W1$ respectively.

Also for $W \geqslant W2$, the adult weight, DW/DT, represents the material added for reproduction over a period of J days. If this total carbon is S, at the end of J days a new cohort is started by taking $Z0 = X \cdot S/W0$, where $W0$ is the initial weight. This process simulates the relation between numbers of eggs produced and amount of food available to the adults, but it neglects the relatively continuous production of eggs which occurs naturally with *Calanus* (Marshall and Orr, 1955). Such artificiality is necessary to keep the population as a single cohort. It is equivalent to assuming that all eggs are produced at, say, $J/2$ days and become nauplii ready to start feeding at J days. The remaining adults at J days are usually few in number and have little effect on the biomass. Instead of merely being "destroyed" at this point, they are preyed to extinction by taking $Z1 = W1 = 0$.

Although this food chain has been kept excessively simple, the number of parameters has proliferated. It seems useful to list these as an indication of how many assumptions have had to be made, and of how many constants have to be given numbers before a computer run can be made.

Variables – T, R, P, W, Z, S.

Constants – V (mixing); A, B (nutrient uptake); C, D, $P1$ (zooplankton grazing); E, F (zooplankton metabolism); G, GX, H, $Z1$, $W1$ (zooplankton mortality); U (zooplankton excretion).

Initial Conditions – $R0$, $P0$, $W0$, $Z0$.

Thus, counting constants and initial conditions, eighteen numbers have to be chosen for each run. If we consider these as degrees of freedom, it would be surprising if some amount of realism were *not* achieved. The problem for this simulation, and for further experimental and observational work, is exactly how these numbers can be defined.

6

Results of the Simulation Model

The main purpose in setting up this simulation is to compare two different "pictures" of copepod dynamics–one derived from the simplest set of assumptions on feeding, respiration, and predation rates, and the other from more complex responses suggested by recent research. These are analogous to the invertebrate and vertebrate types of response described earlier. At the same time, other aspects such as rate of nutrient uptake, copepod reproduction, and predatory pattern will be considered.

[The units of measurement given in the subsequent text figures are those most commonly used in presenting oceanographic data: nitrogen (μg at N/1), particulate plant carbon (mg/m^3), individual copepod biomass (μg C), copepod numbers (numbers /m^2), and population biomass (g C/m^2). For simplicity I refer to copepod "weights," although the values are expressed in carbon units. Usually the ratio of carbon to dry weight is about 0.4.]

Since relatively little is known about some of the parameters, it is easiest to begin by using a set of values which provide a reasonable fit and then testing the effects of changes in these parameters. The output which will be used as a basis for these comparisons is shown in Fig. 6.1.

Basic Simulation

The reasons for the choice of each parameter will be discussed in detail in the remainder of this chapter. Figure 6.1 is presented here to

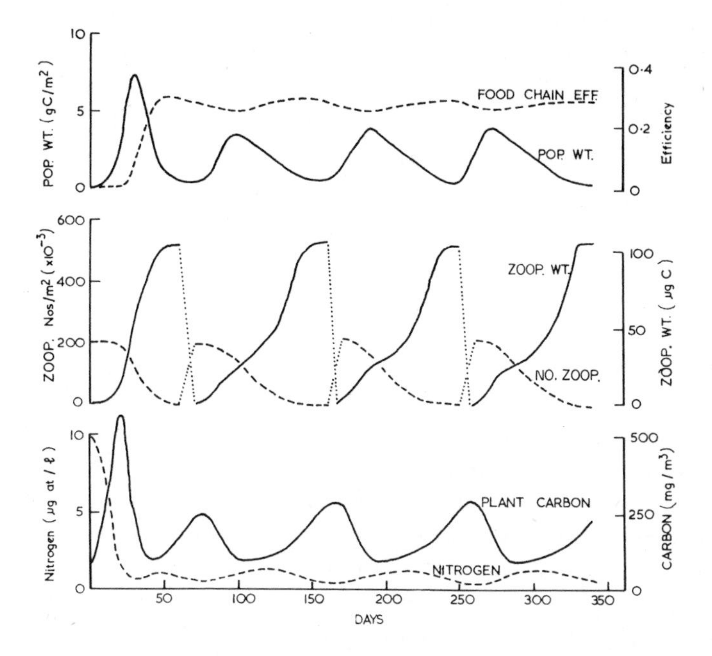

FIGURE 6.1 The simulation to be used as a basis for studying the effects of variations in parameters.

give a picture of the type of results produced by the model. The nutrient and plant carbon concentrations are those calculated to occur in the euphotic zone, the upper 40 m of the water column. The curve of zooplankton weight (in carbon units) shows the growth of an individual copepod, and the dotted lines indicate the points in time at which a new generation is started. The zooplankton numbers display the pattern of mortality. In the upper graph, the solid line is derived from the product of copepod numbers and individual weights to give the total herbivore population weight (again expressed as carbon) under a square meter of surface. Finally, the food chain efficiency at any time is the ratio of total herbivore carbon eaten to plant carbon produced up to that time. The value is very low initially, since there is a lag in grazing compared with plant production and it is this which gives the early peak in plant carbon. The system is run for the unrealistically long time of 360 days to indicate the degree of stability achieved. To obtain this fit, the most important factors in the selection of parameters are as follows: (*a*) a nonzero value was used for $P1$, (*b*) the constant respiration factor, F, was taken as zero so that respiration rate followed feeding rate, and (*c*) the proportional predation factor, GX, was taken as zero and all predation was dependent on biomass (WZ). In other words, in each case the more complex response was used and the simpler type ignored.

In the northern North Sea the euphotic zone is about 40 m deep; the mixing rate through this level is about $V = 0.005$ in the center of this area (Fladen Ground) and increases to about 0.02 in the shallower inshore areas (Steele, 1958). The value used here is $V = 0.01$. For the proportion of nutrient excretion in the upper mixed layer $U = 0.4$, and this would result from equal release at all depths in a 100-meter water column. Changes in V from 0.005 to 0.02, and in U from 0.1 to 1.0, give moderate increases in production and in zooplankton, but do not alter the character of the system. They will be kept constant in the remainder of this section. The plant production of carbon in two hundred days, about the length of the productive season in the North Sea, is 86 g C/m^2, which is of the right order.

In the ensuing discussion of the effects of varying the parameters, U and V and the values for production at the different trophic levels will not be considered in detail, since these are not relevant to the main problem of stability of the system.

Comparison with Observations

The main feature of Fig. 6.1 is the large amplitude of the cycles in plant and zooplankton carbon. The peak value of plant carbon in the spring outburst is observed in the North Sea (Steele and Baird, 1962) and the maximum value of zooplankton carbon, 7 g/m^2, is also reasonable since values of 4 to 12 g C/m^2 are commonly observed. One method of decreasing the amplitude of the cycles after the spring outburst is to introduce a second cohort of zooplankton, as in Fig. 6.2. This smooths the cycling considerably and, presumably, the use of further cohorts could decrease the amplitude even more.

There are, however, two difficulties in attempting such an elaboration of the model. First, although we know in a general way that suc-

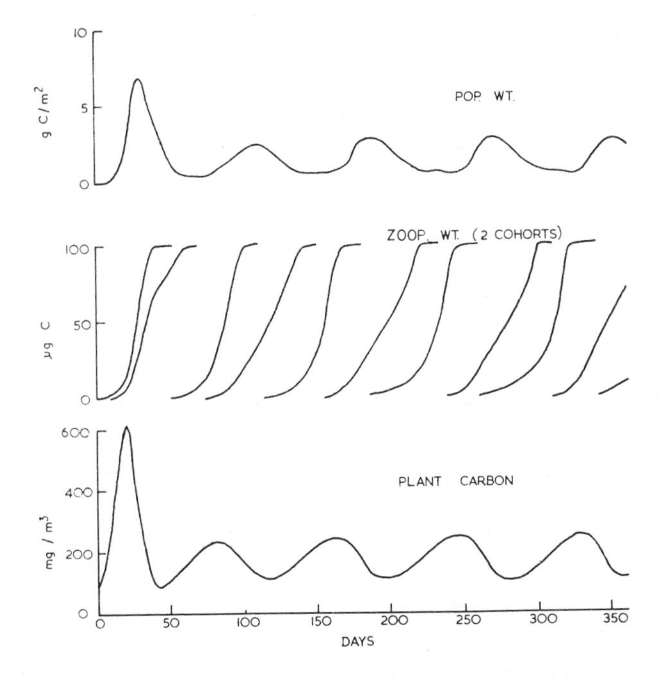

FIGURE 6.2 A second cohort of zooplankton is introduced into the simulation model ten days after the first. Each cohort is half the size of that in Fig. 6.1.

cessive cohorts could arise from occasional increases in phytoplankton in early spring (Marshall and Orr, 1955), we cannot yet quantify the process. More important, none of the observations available are in sufficient detail to provide a critical test of such elaborate models. The range of values of zooplankton biomass in the North Sea during summer (Fig. 4.1) is rather wider than that given by the model. Similarly for plant carbon, the summer range from observations in the northern North Sea–100 to 400 g C/m^2 (Steele and Baird, 1965)–covers that found in the model. Such observations merely show that the model is providing the right scale of values and do not test the details generated by the simulation.

When we turn to the intensive surveys described in Chapter 4 it is possible to compare pairs of plant and copepod observations with the theoretical relation if we use a conversion factor from chlorophyll *a* concentration to organic carbon concentration. From extensive measurements of carbon and chlorophyll in this area (Steele and Baird, 1965) the regression of carbon on chlorophyll gives a ratio of approximately 100:1 at this time of year. A few measurements of carbon during the first survey suggested that the ratio, on the occasion of these surveys, might have been higher and close to 150:1. These conversion factors are indirect methods of estimating plant carbon and indicate a further problem in the testing of models against field observations.

The former ratio has been used in Fig. 6.3 to plot plant carbon concentration in the euphotic zone against the carbon content of the copepod population. Fig. 6.3 also shows the time sequence of plant vs herbivore carbon derived from the model for two values of the threshold feeding level $P1$. The higher value, $P1 = 75$ mg/m^3, is that used in the basic run in Fig. 6.1; the use of a lower value, $P1 = 25$ mg/m^3 (see Fig. 6.5), shows that by varying this parameter the majority of observations can be contained in an envelope of values of $P1$. On the other hand, by using a ratio of 150:1 for carbon: chlorophyll, the original choice of $P1 = 75$ mg/m^3 would give a reasonable fit within the wide limits produced by the scatter of the data.

This juggling with conversion factors and parameters from the model shows that some similarity between theory and observation can be achieved. It should be pointed out that the theoretical population spends about three-quarters of its time on the left-hand side of the curve, so that the distribution of values could be expected to lie in this sector.

While this comparison is in part a test of the model, it is mainly an illustration of the difficulties in getting suitable results for a more elaborate verification. It must be stressed once more that the comparison made here is between a theoretical time sequence and observations collected over an area of the sea in a short space of time. It is possible that the data show varying phases of the cycle occurring at different places. However, spatial patchiness may be causally related to the time cycling through processes of migration of herbivores and carnivores. A properly designed set of observations would need to run over at least fifty days and take account of other factors such as advection and turbulent diffusion of the water. This example emphasizes the fact that small changes in the response of the model are of

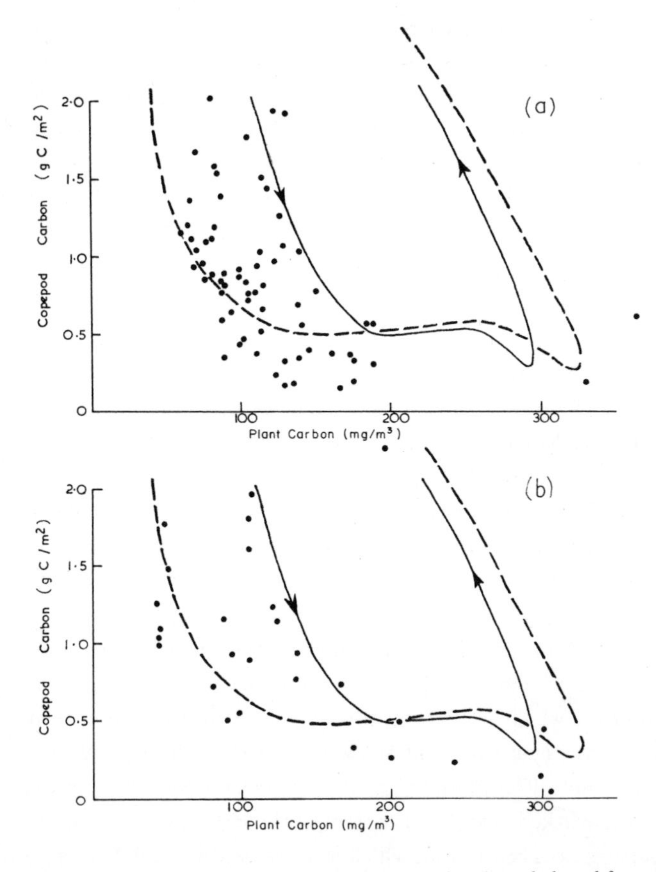

FIGURE 6.3 A comparison of plant and copepod carbon deduced from Fig. 4.6, with the limit cycles for *P*1 set at 75 (——) and 25 (- - -) (two surveys, *a* and *b*).

little interest, since our data are not sufficiently comprehensive to test them. The use of only one cohort in subsequent runs of the model should be adequate to indicate gross changes when different parameters are varied.

Nutrient Kinetics

Variations in the rate of nutrient uptake can conveniently be considered as two separate effects, one when nitrate concentrations are relatively high and one when they are low. The former is only important in the present model during the early period of the spring outburst, and the rate probably depends more on physical conditions than on uptake kinetics. As discussed earlier, initial conditions of mixed layer depth, incident light, and mixing rate may be highly variable. The coefficients used in Fig. 6.1 ($A = 0.2$, $B = 8.0$) give an initial reproduction rate of 16 percent. By varying A as in Fig. 6.4, initial rates of 24 percent and 8 percent are obtained, the former being a relatively high natural rate. The two sets of responses are initially very different, the former producing a large peak in plant and animal carbon, the latter changing much more gradually with no real outburst. The former situation is similar to conditions to the east of the northern North Sea, where water of lower salinity near the surface can enhance the formation of a stable layer, giving high rates of production within this layer and so simulating the theoretical conditions. On the other hand, for inshore areas near the coast of Scotland where mixing is greater, there is often no real evidence of a marked spring outburst.

Thus the model is able to accommodate variations by a factor of 3 in the initial nutrient uptake rate and produce cycles of the other components which are not unrealistic. Especially after the first hundred days the cycles of plant and animal carbon, although of different phase, do not differ in mean values, nor does the food chain efficiency show any marked change. Variations in the coefficient A do affect the nitrogen concentration, which is unrealistically high with a low A value. However, at low nitrogen concentrations it is the value of B which is most important. The value of B used here corresponds to a half-saturation value of 2.7. Reducing this to half decreased the nitrogen concentration to an average of 0.5 μg at N/1, which is not unreasonable, but does not alter the other parameters significantly.

Although it is essential to have a relation between nitrogen concentration and rate of nutrient uptake, the exact form of this relation is not important so far as the stability of the system is concerned. However, the rate of nutrient uptake as a general function of light and mixed layer depth, as well as nutrient concentration, is very important during the spring period. Figure 6.4 illustrates the marked effects on

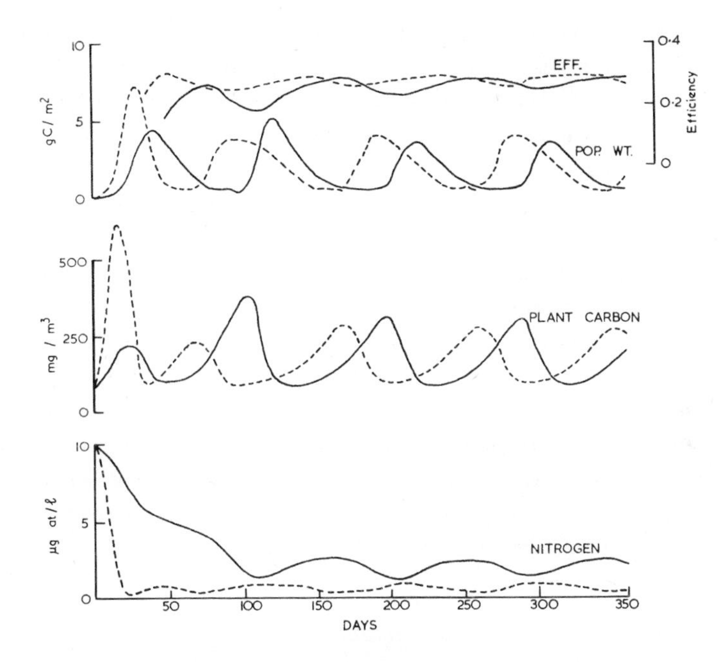

FIGURE 6.4 The effect on the simulation model of varying the rate of nitrogen uptake by the plants (—— coefficient $A = 0.10$, - - - $A = 0.30$).

the zooplankton cycle which can occur during the first hundred days. The differences are likely to be highly significant for certain predators.

Many fish including important commercial species such as cod, haddock, whiting, and plaice have a reproductive cycle which is timed so that eggs, produced before the start of the spring outburst, develop to the end of the yolk-sac stage during the period when copepods are normally plentiful (Jones, 1973). These larvae are not a major predator on the adult copepods, but their survival may depend critically on the numbers and age structure of the copepod population. Jones has suggested that variations in larval mortality consequent on the variations in their food during this period could produce the fluctuations in recruitment of these fish populations, which are very marked and which form one of the main problems in the rational exploitation of stocks. Although the timing of egg production by fish species may be based on the long-term average conditions of plankton blooms in any area, this egg production takes place before the bloom has begun and the timing cannot be altered to take account of variations in the sequence of events during the bloom. The model shows how large these variations can be and suggests that an understanding of the fluctuations in the food of the larvae will depend on a knowledge of the details of day-to-day changes in physical, chemical, and biological parameters.

It is likely that the underlying factors are physical, deriving ultimately from general meteorological factors, such as incident radiation and wind strength, which can determine the sequence and structure of thermocline formation. These factors could be expected to produce effects on production cycles over quite large areas but, as the model indicates, the interactions of the physical and biological processes are extremely complex and the consequences are unlikely to be demonstrated by simple correlations of stock size with one or two physical parameters.

Zooplankton Grazing

The values for C and D which define feeding rate were chosen so that three generations were produced in 200 to 240 days, which is about the period (April-October) during which these generations occur in the North Sea. For these values ($C = 1.6$, $D = 4.0$), the volume of water swept clear of food per day (V.S.C.) reaches the maximum of most experimental values. For a 100 μg C animal (stage V or adult *Calanus*) the relation of V.S.C. and of food removed is shown in Fig. 5.4. The decreasing slope of the feeding curve with increasing food concentration produces a maximum value of V.S.C. and a decrease at higher food concentrations. Such a decrease is found experimentally (see, for example, Mullin, 1963). The relation of theoretical rates to actual behavior is complicated by the vertical migration of the zooplankton, since they may spend only half the day in the upper layer. Thus the actual rate of V.S.C. could be greater than 6 ml/hour. In the few experiments on selective feeding (Harvey, 1937; Mullin, 1963; Richman and Rogers, 1969) rates of 7 to 10 ml/hour have been obtained for *Calanus* on particular phytoplankton species.

The rates required here, although high, are within the range of experimental results as long as one interprets the V.S.C. in terms of selective feeding rather than as a simple process of filtration. The excessive simplicity of the present model is thereby emphasized, since the threshold value, $P1$, may be in part a function of the sizes or species of phytoplankton cells in relation to the size of the individual copepods at the time. Recent experimental work with *Calanus helgolandicus* feeding on cultures (Paffenhofer, 1971) has given even higher filtration rates–200 to 500 ml/day for a 100 μg C animal. These results illustrate the consequences of improved experimental technique, which is likely to be a continuing process. I have not incorporated Paffenhofer's values into the model since, although these and other new data might improve the quantitative validity of the model, I do not think they alter the general conclusions.

The value of $P1$ has been set at 75 μg C/l on the basis of experimental data (Adams and Steele, 1966; Parsons et al., 1969). When this value is reduced to 25 μg C/l (Fig. 6.5), there are distortions in the zooplankton growth curves and a larger amplitude in the plant carbon cycle, but the model retains its essential features and the food chain efficiency does not change greatly. However, when $P1 = 0$ (Fig. 6.5), the system collapses. After the initial bloom, the plants are grazed to a value very close to extinction and remain at this concentration until the zooplankton population has been reduced to an extremely low level. During this period there is no change in weight of the zooplankton, which is a consequence of setting $F = 0$ and assuming that metabolism is realted to feeding. The results demonstrate the need for a threshold even though the actual value can lie within fairly wide limits.

Zooplankton Respiration

The assumption used in the basic run (Fig. 6.1) is that respiratory rate is directly proportional to rate of food intake. Except for adults, this leads to a fixed ratio of growth to food intake expressed as $(0.7\,C - E)/C$. For the values of the parameters used here ($E = 0.5$, $C = 1.6$) this ratio, the "gross growth efficiency," is 0.39, which is sufficiently close to the general average of 0.34 from experimental

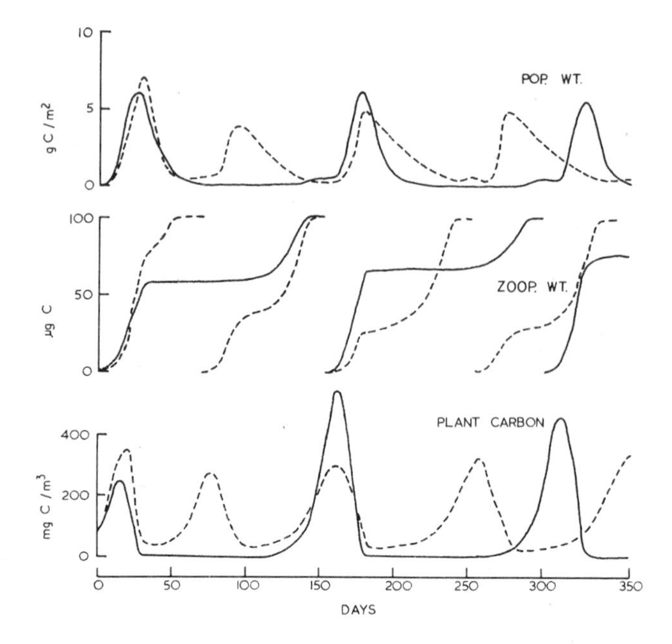

FIGURE 6.5 The output of the simulation model when threshold value $P1$ is one-third its value in Fig. 6.1 (- - -), and when $P1 = 0$ (——).

work (Butler, Corner, and Marshall, 1969). At a food concentration of 150 mg C/m^3 this gives a daily carbon requirement for metabolism of 3 percent of the body carbon for a 100 μC copepod, but at a food level of 500 mg C/m^3 the value is 8 percent. The former is close to values from laboratory experiments (Marshall and Orr, 1955; Mullin and Brooks, 1970b); the latter would appear to be too high. On the other hand, the theoretical consequences of assuming a fixed respiration rate independent of feeding rate can be shown by taking $E = 0$ and $F = 0.21$, equivalent to a 4 percent daily respiratory carbon requirement for a 100 μg C animal. The resulting growth pattern is highly unrealistic (see Fig. 6.6). The choice between these two assumptions depends on the interpretation of experiments on copepod respiration rate.

The problem concerns technical difficulties in the experimental measurement of oxygen uptake. To get a measurable oxygen decrease in

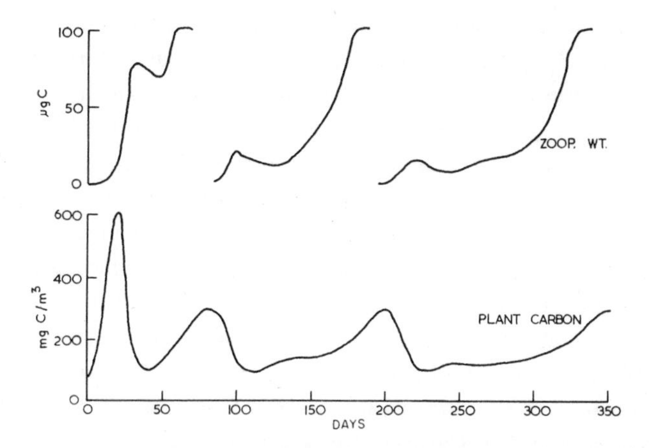

FIGURE 6.6 The effect of changing from a respiration rate dependent on the rate of feeding to one that is independent of it.

a closed bottle, the concentration of copepods has to be ten to one hundred times the natural concentration in the sea. Even if natural seawater containing some plant material is used for the medium, this food is very rapidly reduced to a low level and the copepods stop feeding. Very often, however, filtered seawater is used as the medium and no feeding is possible. Further, the animals are usually kept for 12 to 24 hours since "immediately after capture *Calanus* has a considerably higher respiration than some hours later" (Marshall and Orr, 1955) and respiration continues to fall for several days thereafter (Marshall and Orr, 1958). Although some effect of lack of food is usually acknowledged, the main factor is taken to be acclimation after the stress of capture. The alternative interpretation is that the initial rates are those of actively feeding animals and the decrease is the result of a comparable decrease, or absence, of food.

Some experiments have been performed to illustrate these effects. The animals used were taken from natural populations occurring in coastal waters and consisted of a mixture of *Temora longicornis* and *Acartia clausi*. Approximately five hundred animals were kept in one-liter bottles. The water in the bottles could be exchanged through fine meshes which permitted a renewal of the food without loss of animals. In the first experiment two plant concentrations were used; natural seawater, and a higher food level produced by enriching the natural seawater with nutrients and allowing it to stand in the light for three days before use. During the experiment the food supply was renewed each day for three days and the oxygen consumption followed in the intervening periods using oxygen electrodes. The treatments were in duplicate, and there were controls without animals. The decreasing rate of oxygen uptake (Fig. 6.7) is an indication of adaptation to food supply.

In a second experiment (Fig. 6.8), two pairs of bottles were used to study the effects of infrequent flushing resulting in a low food level in the bottles, compared with an increased frequency of flushing after the first day. The results show that the usual decrease in respiration can be reversed by flushing and, in fact, with frequent flushing the respiration rate was higher at the end of the period for those bottles with frequent renewal of food. In this experiment the changes in chlorophyll *a* in the bottles were used to provide an index of filtering rate, and there was a significant positive correlation (at the 5 percent level) between respiration and filtering rates. Within the limitations of this kind of experiment, the correlation provides a test of the prediction from the model that metabolic rate must be dependent on feeding rate.

It is unlikely that there is no "basal" metabolism, or that the relation to food concentration is as exact as that used in the model. There is no point in trying to go beyond this theoretically, since a greater degree of simulation is not possible for copepods on the basis of existing data. For example, the metabolic response to a change in food concentration may operate on various time scales, part of it occurring in a few hours, but some components may take days or weeks.

Predation

For predation on the copepods the simulation in Fig. 6.1 contains two thresholds, both of which theoretically will tend to stabilize the whole system. The removal of the threshold on weight ($W1 = 5\ \mu g$ C)

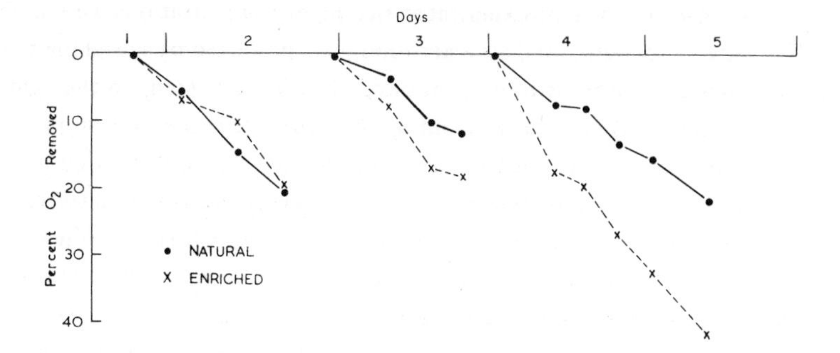

FIGURE 6.7 The rate of oxygen uptake by copepods (approximately 500/1) in seawater with natural and enriched phytoplankton populations (food supply renewed every 26 hours).

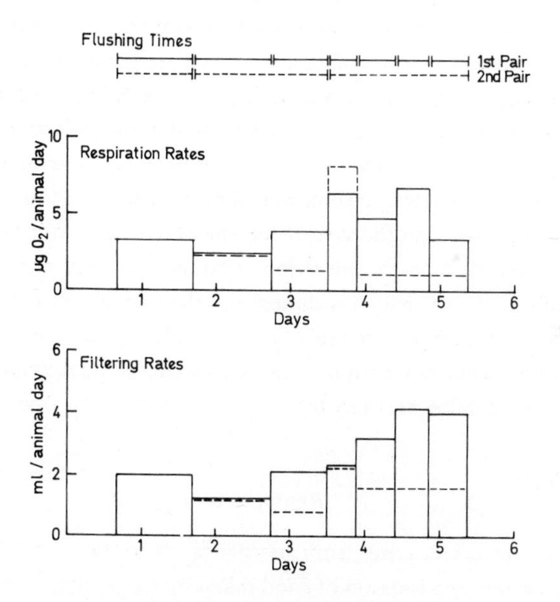

FIGURE 6.8 Respiration and filtering rates of copepods. The double vertical lines indicate when the bottles containing the copepods were flushed with seawater to introduce a fresh food supply. The respiration rates were estimated from oxygen decreases and are given for the time intervals for which the filtering rates were determined from chlorophyll decreases.

has little effect, but the removal of both thresholds (Fig. 6.9) produces a much more variable response than that of Fig. 6.1. The system is, however, still "realistic" within the wide limits already mentioned, although it is likely that it would respond much more erratically to stochastic fluctuations. While these thresholds do have a stabilizing effect on the whole system, their removal, giving a basically unstable prey-predator relation, does not cause the complete breakdown which occurred when the copepod grazing threshold was removed. This supports the conclusions of the earlier chapters: that such thresholds anywhere in the system can be significant in determining the level of stability, but that certain of them may be most critical, since without them the whole system breaks down.

The predation rate is defined in terms of a response to herbivore numbers and size rather than as the behavior of specific predators; therefore the effects of a different rate of predation are simulated by altering the coefficient G. A doubling of G, as in Fig. 6.10, has a comparatively minor effect, but a low predation rate completely alters the pattern by an overproduction of juveniles which keep their food only just above the threshold concentration and thus grow very slowly. Even though the system remains stable, it is unrealistic because of the long life of each generation. Also the herbivore population level is about 6 g C/m^2, which is three times the average in the North Sea. Either the predation rate must be reasonably high or there must be some extra control of population exerted on the population at the time the juveniles start feeding.

Reproduction

This same effect arises from another factor in the model. The symbol X defines the fraction of food taken by the adults, which forms the biomass of the juveniles of the next cohort. I have suggested that the juveniles contain about 30 percent of the carbon used in egg production. To allow for male feeding and for some mortality, the actual value of X used is 0.1 rather than 0.3, and this has been kept fixed up to now. The device is fairly arbitrary, since so little is known of the processes involved. The effects of variation in X were studied by run-

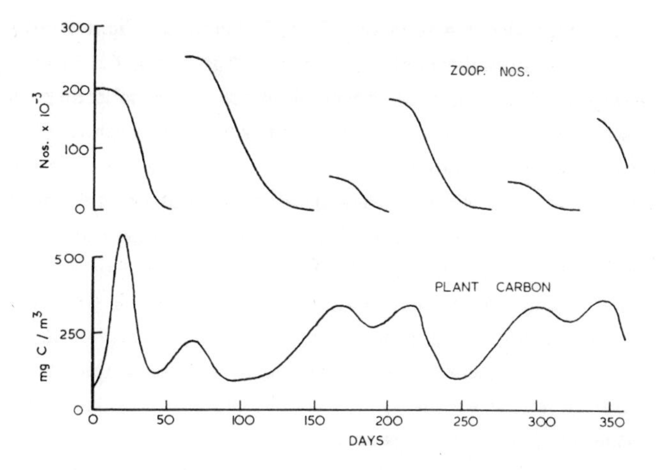

FIGURE 6.9 The effect on the simulation model of removal of both thresholds for the predators ($W1 = Z1 = 0$).

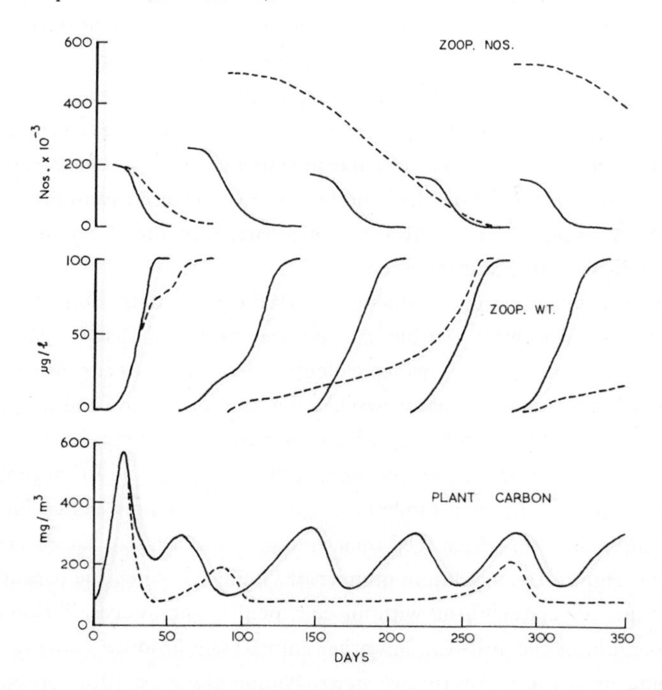

FIGURE 6.10 The effect of the simulation model of variation in the overall rate of grazing from $G = 0.02$ (- - - $G = 0.01$; —— $G = 0.04$).

ning the model with $X = 0.03$ and $X = 0.3$. The results were almost identical with those obtained by varying G (Fig. 6.10). Thus the original simulation (Fig. 6.1) depended on the "correct" choice of two parameters, G and X, but in neither instance is there any evidence on which to base the choice.

This defect can be cured mathematically by making the juvenile/egg ratio dependent on the amount of carbon used for egg production. Instead of $Z0 = X \cdot S/W0$, take $Z0 = X \cdot S/(W0[1 + S])$. This means that, instead of having egg production directly proportional to the food available to the adults, an upper limit is set to the number of eggs which can be produced by the adult population. This simple mathematical trick removes nearly all the aberrant features associated with low values of G or high values of X.

The trouble with this kind of assumption is that it does not help in understanding the processes involved. Further, it operates in terms of population numbers rather than individual behavior. It is possible to produce hypothetical mechanisms. For example: the range of carbon concentrations of eggs is 10 to 100 mg C/m^3 if all eggs occur in the euphotic layer. Since the eggs are about 150 μ in diameter, they are similar in size to large phytoplankton cells and "hungry *Calanus* will eat their own eggs" (Marshall and Orr, 1955). Selective cannibalism above a certain egg concentration, therefore, is an adequate, but highly speculative, means of control.

This problem has direct analogies with those of population control in terrestrial animals. For birds, it has been proposed (Lack, 1966) that the main mortality, which is density dependent, occurs outside the breeding season, usually associated with shortage of food during the winter. Beyond this, Lack also concludes that the reproductive rates of birds have evolved through natural selection and so, in general, are as rapid as the environment and the birds' capacities allow. For example, in birds which feed their young, the clutch size has evolved in relation to the maximum number of young for which the parents can provide enough food without detriment to themselves. Within this general limit the birds (or any other animal) will produce as many young as possible. Alternative views (Wynne-Edwards, 1962) lay more stress on control of population through numbers of young produced and less emphasis on later mortality.

Much of the difficulty lies in the theoretical definition and experimental proof of "density dependence." In this simulation of a copepod population I have shown that a stabilizing type of predation is not essential, at least for a deterministic model. On the other hand, the rate of predation on the larger animals is important but, within the definition of this system, variations in this rate have essentially the same effect as differences in mortality (or production) of eggs. A theoretical approach cannot settle this question. It is possible that both types of response will occur together, or separately, to meet different environmental fluctuations. Probably only experimental work, rather than more observations, can answer such questions.

There is a third "solution" to the problem of an excess of juveniles: dispersion. This is important and probably essential for terrestrial animals where breeding occurs in restricted areas. Dispersion on land, or in the sea, could overcome the effects of a high density of juveniles occurring only in one area. The assumption in the previous discussion was that the excess populations were sufficiently widespread for dispersion not to be an adequate solution. Since dispersion in the sea is probably a more important process than on land, I shall consider its role further in the next chapter.

Efficiency

Returning to Fig. 6.1, the food chain efficiency rapidly reaches a relatively steady level of about 27 percent. This is 12 percent lower than the gross growth efficiency of 39 percent. However, of the 86 g C/m^2 produced in 200 days, 14 g C/m^2 were mixed out of the upper layer and thus were not eaten (within the terms of this model). The ecological efficiency (zooplankton predation/zooplankton food) accordingly is 32 percent. If production stopped at the two hundredth day and nearly all the remaining crop of zooplankton on that day were eaten, the ecological efficiency would rise to 37 percent. The remaining 2 percent is accounted for by the energy losses and "mortality" of eggs. This value is so small that in terms of energy conversion, rather than population control, the exact definition of metabolic conversion of eggs to nauplii is unimportant. The implication is that copepod populations expend very little energy in ensuring that their offspring survive to form the next generation. It is for this reason, again within the terms of the model, that the ecological efficiency is so high in relation to growth efficiency. The value obtained would fit with the estimated value required for the food web in the North Sea (Chapter 2).

At each stage in the development of this model there was a choice between a relatively simple and a more complex hypothesis of the copepod's behavior or of the interaction with other trophic levels. The basic model, Fig. 6.1, used the complex hypotheses, and it has been shown that in each case where the simpler assumption is used, the model tends to be less realistic in appearance. As a final comparison, the three simple assumptions have been taken together: no threshold grazing concentration, a fixed respiration rate, a proportional predation rate. With these conditions the only "safe" response is one with a very large predation rate on the copepods.

Figure 6.11 shows that even a moderately high predation rate ($GX = 0.15$) is insufficient, since after the spring outburst the second generation of copepods is able to eliminate all the food. By putting the rate higher ($GX = 0.20$), the copepods are always at a low level and a consistently high plant food supply is ensured. As a consequence, nearly all the phytoplankton is mixed out of the upper layer and can be assumed to end up on the bottom as part of a different, decomposer, ecosystem. A further consequence is that the food chain efficiency of the herbivores is very low, barely reaching the 10 percent level after 360 days. This response certainly indicates the need for some set of more complex responses to obtain anything resembling a marine ecosystem, even if it does not prove that those used here are of the correct type.

It is tempting to compare the picture given by Fig. 6.11 with the general hypotheses about terrestrial ecosystems proposed by Hairston, Smith, and Slobodkin (quoted in Chapter 1). The relative levels of Fig. 6.11 are in principle those of a terrestrial environment and so would tend to support the idea that herbivores on land are not capable of exercising direct control over their food supply.

Conclusions

The model has been used in three ways. It is intended to give a rough simulation of conditions in the North Sea. This has not been stressed, partly because of the inadequacies of the data, but mainly because the introduction of more parameters (such as temperature, mixed layer depth, or incident radiation) might obscure the discussion of variations in the behavioral type of responses. It has been pointed

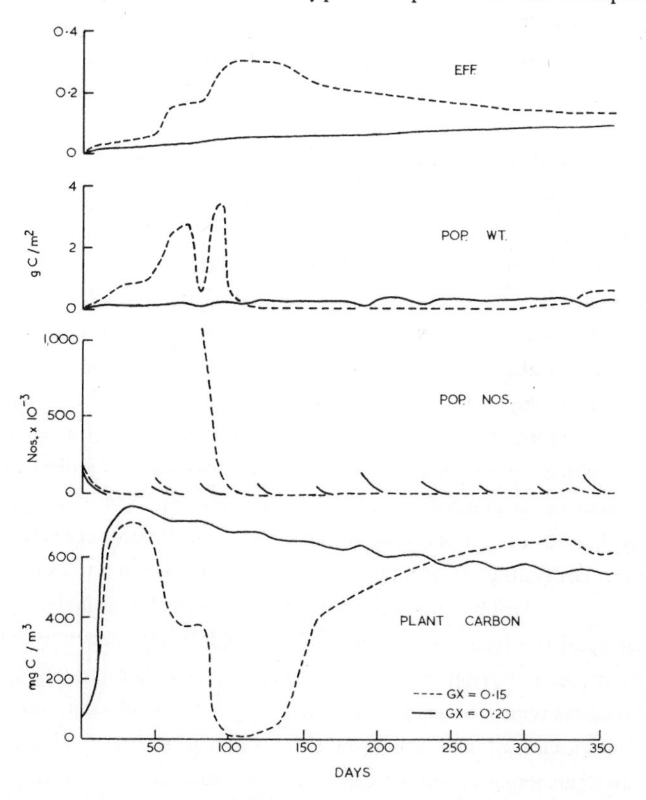

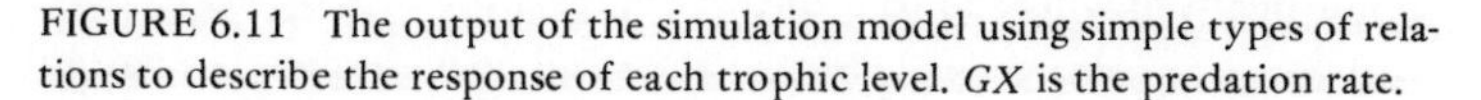

FIGURE 6.11 The output of the simulation model using simple types of relations to describe the response of each trophic level. *GX* is the predation rate.

out that the omission of plankton patchiness severely limits the validity of comparisons with observations. This type of model could be appropriate to areas where there are large plankton blooms in spring; it is unlikely to apply directly to open tropical waters characterized by the relatively constant levels of plankton biomass. A model simulating conditions in high latitudes has the advantage that it can reflect the large-amplitude perturbations which occur there and with which such an ecosystem has to cope. It is possible that in tropical conditions a quite different set of responses could exist; however, I have tried to show that this could not be based simply on the greater diversity found there. Further, on the edge of such tropical regions, for example in the Sargasso Sea off Bermuda, a predominantly tropical environment has occasional bursts of production (Menzel and Ryther, 1961) and appears to respond in a manner similar to that found in more northerly regions (Steele and Menzel, 1962).

The model also could have stochastic effects incorporated into the time sequences, particularly for the physical variables. This is a more sensitive and realistic way of studying the effects of variation in the parameters. However, if the deterministic model can respond suitably to large changes in a parameter over the period of 360 days, then random fluctuations around the original value of that parameter should not affect the response seriously.

The second use of the model has been to demonstrate the possible function of certain types of behavioral response as control mechanisms for the whole system. Nearly every response used has been density dependent in the general sense that the relationship used is curvilinear. At the lowest level, it is essential that there is a relation between nutrient concentration and rate of nutrient uptake, but the exact form of this relation is not significant for the general stability or productivity of the system—although, of course, it does determine the nutrient concentration in the upper layer. At the other end, the system survives with a basically unstable prey-predator relation, but is much improved when a stabilizing threshold is introduced. This illustrates the general thesis of Chapter 3 that such relations at any point in a web will tend to improve the response of a model to perturbations; a priori it is likely that they will be widespread in natural systems. However, their occurrence does not mean that they can produce the type of system which is found naturally; there are still points in a web where the existence of such mechanisms is most critical. In this model the interaction of zooplankton and phytoplankton appears to be one of these critical points. The exact shape of the relation and the behavioral basis proposed for it may be too specific.

It is one of the advantages, as well as a disadvantage, of a formal mathematical relation, that it can be given a wide range of natural interpretation. In this case the conclusion could be that a relation of this general *type* is required. Further, the control used has been based on only one activity, feeding behavior. It has been shown that this is insufficient, since the very simple formula used for reproductive rate is inadequate to account for all situations. Again, a more complex relation could have been introduced at the outset for reproduction. It seemed more appropriate to use this feature as a means of stressing the artificiality of this (or any) model, and of the interaction between theory and experiment which is an essential part of the process. Comparisons with similar problems in terrestrial ecology, pertaining to the relative importance of control during and outside the breeding periods, are also highlighted.

The third and most speculative use of the model has been to suggest comparisons with terrestrial systems. As already stated, the existence of a density-dependent relation does not in itself presume a critical role for that interaction in terms of the whole system. One would expect such relations to be widespread in terrestrial environments and to vary according to the particular type of environment. Yet it is still possible to propose as Hairston and his associates, Lack, and others have done that, because food of predators is usually limiting whereas that of herbivores (as a class) is not, then the way in which control operates in the former relation is critical for the whole system. Even though there are examples of density dependence at the terrestrial herbivore level, these may be more of a fail-safe operation than a normal type of control.

It must be stressed that this marine-terrestrial dichotomy is not absolute. The macrophytes on rocky shores are probably best considered as "terrestrial," since nearly all seaweed produced enters a decomposer type of system. On the other hand, arctic lemming cycles are almost "marine," since the guiding hypothesis according to Pitelka (1964) is that these cycles are "a result of interaction between herbivore and vegetation mediated by factors of nutrient recovery and availability in the soil." It is tempting, but presumptuous, to suggest that in such marginal environments with very limited plant food, terrestrial herbivores do not have the quality of control of their marine counterparts.

7

Speculations

In any system on land or in the sea, nearly all the biological energy produced by photosynthesis is eventually utilized by heterotrophs as food (Hutchinson, 1948). Our practical concern is with that part which passes through us. In the management of our resources a question of increasing importance is the potential yield from the sea compared with that from land. An indication of the general levels in a "developed" country, the United States (Fig. 7.1), shows how little plant production ends up as animal products used by man. In the North Sea, our yield appears to depend upon nearly all the plant production entering chains which end at commercially harvestable species. Although there is much more management on land, with shorter and simpler food chains than in the sea, the yield of animal matter in the United States is 0.5 percent of the plant production, compared with 0.8 percent in the North Sea. This is probably an unfair comparison, as the North Sea is a relatively rich and heavily exploited area. Such areas of shallow water form a comparatively small proportion of the world seas, and our knowledge of potential yield from the open ocean is slight indeed.

Ryther (1969) has produced a very generalized budget for the world oceans, which is given in Table 7.1. The high yield from upwelling areas depends on the harvest of fish, such as anchovy, which are predominantly herbivorous. The ratio of yield to primary production for coastal regions is in rough agreement with the North Sea data. It is the oceanic values which are most speculative. According to Ryther, the long food chain arises mainly from the fact that the majority of

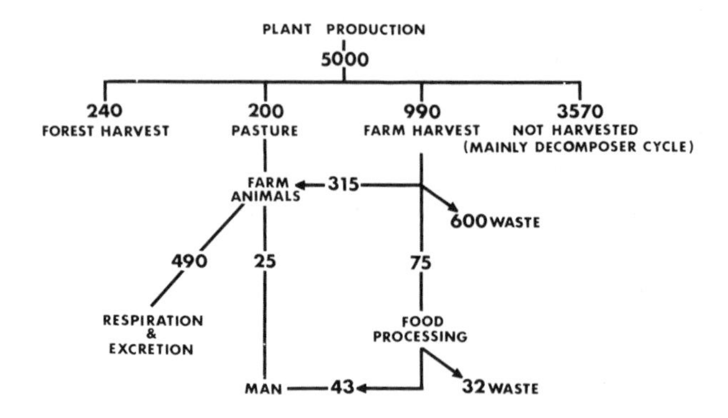

FIGURE 7.1 The flow of plant production in the United States; units are 10^6 tons dry organic matter/year (from Kneese, Ayres, and d'Arge, 1970).

phytoplankton cells in such areas are very much smaller than coastal or upwelling species and an additional step accordingly is introduced into the food chain, the microzooplankton. Also at the upper levels of the web the yield is normally in large fish such as tuna; hence the assumption of five trophic levels. Further, as discussed earlier, the higher trophic levels may have lower ecological efficiencies, so that the average values used in Table 7.1 are much lower for oceanic than for upwelling areas.

At the end of Chapter 2 it was shown that for areas like the Sargasso Sea the rate of nutrient input was much lower than the equivalent primary production; in such areas, therefore, long and diverse food chains and a very low yield at higher levels are to be expected. Whether this conclusion applies to all parts of the open ocean has been disputed, but the relevance here of this example is to illustrate that these arguments are based on the premise that effectively all the energy goes through the chain, and that average efficiencies are taken as 10 to 20 percent, much higher than the 0.5 percent for managed herbivores on land (Fig. 7.1). High efficiencies in the sea, deduced for the North Sea, are probably a general feature of many marine environments.

Much of this difference in herbivore efficiency can be explained in terms of the small fraction of terrestrial plant material which is metabolically utilized after ingestion. A further factor is that in commercial

TABLE 7.1 The relation of fish to plant production for types of sea area (from Ryther, 1969).

	Upwelling	Coastal	Oceanic
Percent ocean area (total area 362×10^6 km^2)	0.1	9.9	90
Primary productivity (g C/m^2 year)	300	100	50
Total plant production (10^9 tons C/year)	0.1	3.6	16.3
Number of trophic levels	1.5	3	5
Average ecological efficiency (%)	20	15	10
Fish production (10^6 tons C/year)	12	12	0.2

terms we are concerned with homeotherms on land but poikilotherms in the sea. The latter are usually considered to be capable of much higher production efficiencies than warm-blooded animals (Engelmann, 1966; McNeill and Lawton, 1970). Even so, the values demanded for herbivores in the open sea are very high and even the carnivores have to be near 10 percent efficiency to fit the observed yields.

For most young, actively growing animals (both terrestrial and aquatic) the maximum efficiency of conversion of assimilated food into growth appears to lie between 30 and 40 percent. This may be determined by the efficiency with which protein in food can be converted into body protein. When one considers the yield averaged over many generations from a population to its predators as a percentage of the food assimilated by this population, the values are much lower. The classical value for this ecological efficiency was 10 percent (Slobodkin, 1961). The range now may go from 2 to 5 percent for mammals up to values of 20 to 25 percent for zooplankton.

Since in no case does the ecological efficiency equal the growth efficiency, the difference in large part must be associated with the means whereby an organism ensures the survival of its young. These mechanisms are extremely varied on land, but are based on the need to ensure that the young are in a position to have a good chance of adequate food. The offspring may be inside the mother, inside an egg, in a seed, in a nest, or any combination of these. Nearly all require additional energy beyond the actual feeding process (nest-building, the survival of parents after egg-laying, and so on). Associated with these processes are the forms of behavior usually referred to as territorial activity, where the numbers in breeding groups may be related to food supply (Watson and Moss, 1970). An additional or alternative method occurs through dispersal, but again this involves activity by the adults or preadults. It is generally true that in nearly all species of flying insects the prereproductive adults engage in special dispersal flights during which they fly upward from the ground and are carried along in large numbers by wind currents (Johnson, 1969).

In the sea there is also a variety of methods of reproduction, but the vast majority of species release into the water eggs, or larvae at the yolk-sac stage. There are viviparous fish, species which produce their young in egg cases, and many other examples of similarities to methods used on land, so that alternatives have been evolved. The predominance of release to the upper layers of the sea suggests that this has certain advantages. The most obvious is that the fish larvae, at the end of the yolk-sac stage, are in a position to start feeding on the copepod populations and, as described earlier, part of the success of any larval population depends on the abundance of the right size of copepods at this point in the life of the fish larvae. Once eggs are released into the sea, they are dispersed partly by ocean currents, but mainly by the lateral turbulent diffusion of the water. The important feature of this latter mechanism is that the rate of diffusion increases with the size of the patch being diffused.

Joseph and Sendner (1958) have produced a theoretical description of this process in the absence of land boundaries or barriers within the sea (such as changes in density structure). To give some idea of the scale of the process, I have in Fig. 7.2 superimposed on the northern North Sea the theoretical distribution of the spawning products of one fish (10^6 eggs) after 28 days. The upper values in Fig. 7.2

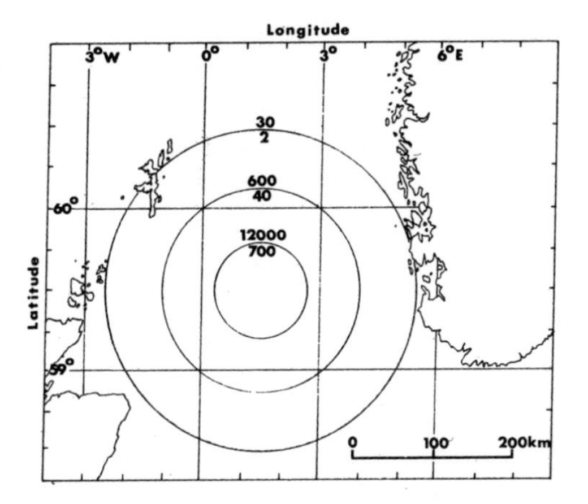

FIGURE 7.2 Theoretical spread of larvae based on Joseph and Sendner (1958) and superimposed on the northern North Sea. Distribution of 10^6 eggs after 28 days: upper numbers, with no mortality; lower numbers, with 10 percent mortality per day (number/km^2).

assume no mortality, the lower a mortality of 10 percent per day, which is about the rate for eggs and larvae of haddock (Saville, 1956). This illustrates the scale of the process and shows that in principle even with a heavy mortality, at least two larvae per square kilometer would reach the post-yolk-sac stage over the whole of the northern North Sea. Further, the intermixing of the water which occurs means that a wide range of type and concentration of food will be available to the surviving larvae compared with the food at the original place of spawning. It is likely that the food supply to the larvae at this period is one critical factor in determining the later adult stock (Jones, 1973).

As a result of dispersal and previous mortality, however, the density of larvae is sufficiently low so that their grazing is not an important part of the predation on the copepods (Rosenthal and Hempel, 1970). As with insects (Clark et al., 1967), dispersal can minimize the probability of intraspecific competition for food. These means of dispersion in the sea are without any energy requirement on the part of the parent or of the larvae. In this sense, part of the energy provided by winds and tides to drive the diffusion process is being utilized by the fish populations. In practice, diffusion does not operate so simply, and spawning populations tend to aggregate in areas where the added effects of currents and shore or water barriers act to limit dispersal to areas smaller than the theoretical distributions (see Fig. 7.2). Presumably it is in this way that localized spawning stocks have evolved to optimize the particular land and sea topography.

Some energy is used by fish in migration to the spawning area and there is a large egg production, of the order of 10^5 to 10^6 per female, which balances the loss of the unprotected eggs and early larvae. For copepods there is no evidence for purposeful migration, the numbers of eggs produced are relatively small (10^2 to 10^3) compared with fish, the time before feeding starts is shorter, and therefore the mortality at this stage is, apparently, much less. Yet the same advantages of dispersion should operate. This process could not be included in the model, but it will need to be considered in more complex theoretical developments. It is relevant to the problems of patchiness discussed in Chapter 4 and affects practical decisions about the planning of sampling as well as concepts of population survival.

In general, then, there is a question of how we view the basic physical processes in the sea. Dispersal could be considered as a purely mechanistic process, not utilized, but simply endured. It may have certain advantages at smaller scales in damping out plankton variations (see Chapter 6). On the other hand, it is a possible but very tentative speculation that turbulent diffusion in the upper layers of the sea, which has no real counterpart on land, can provide a form of "energy" to give relatively high values to ecological efficiency for copepods and certain fish species. At the same time, the random element inherent in such dispersal processes needs to be considered as a factor operating on the control of stability of populations through the effects on recruitment. For fish the general control appears very loose, since there are large fluctuations in the year-to-year strength of the recruiting stock for most species. Murphy (1968) has shown that there is a rough correlation between the amplitude of these fluctuations and the number of years of adult life when spawning can occur. His relation is for pelagic fish and is complicated by the fact that the main short-lived species, anchovy, with only one or two significant spawnings, is a herbivore. For each species of fish a relation of this type is obviously important in terms of population survival. It is especially significant for commercially fished stocks where exploitation tends to reduce the age structure to an artificially young distribution and thereby enhances the probability of a set of poor years of recruitment decreasing the size of the whole stock below a viable level.

On the other hand, the extreme fluctuations in recruitment of species of carnivorous fish—for example, haddock (Fig. 7.3)—when compared to that of birds (Lack, 1966) suggests that, although this is significant for the fish themselves, it may be less important for the rest of the system. Again this could be related to the type and critical nature of the control exerted by predators in the different environments. The comparative regularity of anchovy recruitment could be as much a function of its trophic level as of its short life-span.

It has been suggested here that predation on copepods in the spring, when herring and mackerel are major predators, may operate mainly on the larger stages. This seems most efficient for a predator which is relatively big in relation to the size of its prey. There is some evidence (Steele, 1965b) that when smaller species of copepod replace *Calanus* in the North Sea, the growth efficiency of the herring decreases. A large part of this energy intake in spring goes into fat stores to carry the animal through the reproductive phase in the autumn and then through the following winter. This sequence depends on the existence of an adequate food supply, but the implications of the model and of field data are that the patterns of harvesting are, mechanistically, in terms of efficiency of conversion of energy to a higher trophic level rather than control of population fluctuation.

For the energy flow within the bottom fauna, it is difficult to assess rates at intermediate trophic levels because the quantity of energy going through bacteria is still unknown. Further, the reworking of bottom sediments by the fauna appears to produce a cycle, rather than a link, with the bacteria operating on fecal material to produce new sources of food for the fauna (Newell, 1965; Johannes and Satomi, 1966). In this cycle the fauna rather than the microflora could be rate-limiting (Hargrave, 1970), as has been suggested for the comparable events in soil (Macfadyen, 1961; Edwards and Heath, 1963).

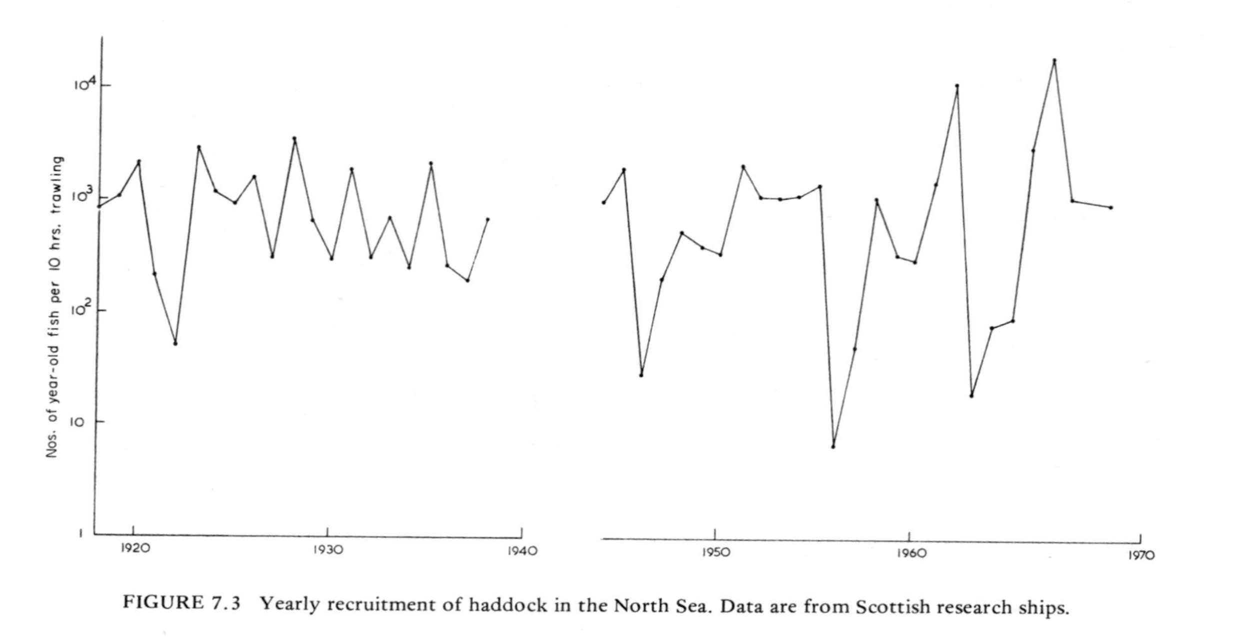

FIGURE 7.3 Yearly recruitment of haddock in the North Sea. Data are from Scottish research ships.

It is not possible to consider the effects these factors might have in producing varying food supplies for the demersal fish. The large fluctuations in annual class strengths of demersal fish have some effect on growth rate (Gulland, 1970), but not as marked as one would expect if their food were exceedingly limited. For small plaice in tank experiments with natural food it has been shown (Edwards, Finlayson, and Steele, 1969) that growth rate is dependent on the relative concentrations of fish and their invertebrate food, rather than on the absolute densities. The same relation appears to hold in terms of the number of young plaice recruiting to an inshore nursery area where these fish occur (Steele et al., 1970). Thus, to some extent, demersal fish may respond to fluctuations in their food supply by variation in recruitment, rather than by changes in growth rate of individual fish.

In turn the fluctuations in quantity of benthos can still depend on events in the supernatant water, since many of the benthic invertebrates have a pelagic larval phase. These larvae, like the fish larvae, may not be highly significant as predators on the plankton; however, they will depend on the density of this plankton for their survival. The benthic fauna, vertebrate and invertebrate, gets its energy from fallout from the pelagic system in a manner comparable to the way that the decomposer system in the soil depends on production in the air above. Beyond this, however, part of the control of the benthos may be a consequence of the degree of regularity in the phytoplankton-zooplankton cycles particularly, in northern latitudes, during the spring bloom. The plant-herbivore-benthos populations in the sea are much more closely linked than the equivalent systems on land.

A possible consequence of coupling between different parts of the web can be seen in the changing yields of pelagic and demersal fish in the North Sea, as shown in Fig. 2.5. There was little change in either component until the early 1960s when the introduction of the highly efficient purse-seine increased the yields of herring and mackerel so markedly that, for herring especially, there are fears that the stock sizes may have dropped to a level so low that recruitment has been affected (ICES, 1970). At the same time there has been a marked increase in the yield of demersal species, although for quite different reasons. This results not from changes in fishing effort, but from the occasional occurrence of exceptionally good recruitment, especially in haddock (Fig. 7.3), but also in other species. The same years, 1962 and 1967, that were particularly fine for haddock were also years of high recruitment for whiting.

One hypothesis is that a removal of predatory herring and mackerel would increase the copepod populations in the spring when they are required as food by the larvae of demersal fish. This could increase the probability of good broods (Jones, 1973). However, plankton samples collected regularly by merchant ships during the period 1949–1969 show a significant decline with time in both copepod numbers and biomass in the North Sea (Glover, Robinson, and Colebrook, 1970). Further, the same decline was found in the northeast Atlantic, where there is insignificant commercial fishing. Thus the evidence is against any direct association of plankton numbers with a fishing effort sufficiently intense to endanger the fish stocks themselves. It is more likely that some of these changes are a result of the relatively long-term climatic variations in the North Atlantic region (Dickson and Lee, 1969).

Another hypothesis is that changes have occurred within the food chain, at levels between the herbivores and the commercially exploited fish species. The elementary theory of Chapter 3 suggests the way in which an alteration in the magnitude of a controlling type of link can take a system across a threshold from a stable to an unstable situation. The analogy between such a simple theory and the increased amplitude in recruitment is extremely tenuous and there is no evidence that changes of the type required occur within the system. Such evidence would be very difficult, if not impossible, to collect by sampling alone. This illustrates both the need for a theoretical framework and the inadequacies of any theory unless it can be supported and then tested by experimental work.

The relative importance of climatic change and of overfishing presents the same problem in the interpretation of events leading to the virtual disappearance of the Pacific sardine. For a similar type of change in the English Channel, where herring were replaced by pilchards (Cushing, 1961), there is no evidence that the alteration can be attributed to the effects of fishing. Recent observations (Russell et al.,

1971) suggest that the earlier changes are being reversed. In both these cases, although there were also variations in the species composition of the plankton and in nutrient cycles, there is no suggestion that the energy flow at the plant-herbivore levels was significantly changed by the alterations in the fish populations. In the Antarctic the virtual elimination of the whales which were a major predator on herbivorous crustacea (krill) does not appear to have caused any massive imbalance in the system. Investigations by the Russians (Anilov et al., 1969) show the expected type of interaction between the phytoplankton and the krill and suggest that, in the absence of whales, direct harvesting of krill is an economic proposition.

All this rather anecdotal evidence on changes in predator populations is complicated by the interaction of long-term natural trends with interference by man. Nor are the commercial species the only predators on the plankton. This has been illustrated by the food web in the North Sea; and also by the existence of significant fish populations which live on krill (Marti, 1967); it is difficult to quantify the changes in predation. The results do indicate that severe pressure by man on pelagic predators such as whales, herring, or sardine affect the survival of the species through inadequate recruitment rather than through any induced changes in the rest of the food web. This can be taken as circumstantial evidence for the inherent stability of the phytoplankton-zooplankton link, and tends to reject the alternative "terrestrial" hypothesis that predator control is essential for lower trophic levels.

In the description of the model no attempt was made to simulate any type of copepod except *Calanus*. The simulation has been run with values for minimum or maximum animal carbon equal to those of smaller copepods such as *Pseudocalanus*, but this required changes in the predation parameters. Further, if one tries to run a "*Calanus*" and a "*Pseudocalanus*" simultaneously, the latter always wins because it reaches maturity earlier. These attempts, apart from indicating the inadequacies of the model, show the need to have several predators when there is more than one prey. It is possible that such a development of the model would give the same result as the analytical two-prey–two-predator system at the end of Chapter 3—that is, that by varying the coefficients one can go from a large prey–large predator to a small prey–small predator through an intermediate stage where all four coexist. This has not been tried yet but, very tentatively, we can say that it would provide a means whereby a gradually changing environment could pass through critical points at which the structure of the system changes completely. At present, however, the analysis given here cannot include the consequences of long-term climatic trends. Nor is it relevant to consider the evolutionary or selective pressures which could "produce" threshold responses that operate in the short term to meet the effects of day-to-day changes in the environment.

If fishing appears to have relatively little effect on the rest of the food web, one other recent consequence of human activity—cultural eutrophication—has caused considerable alteration in large freshwater lakes. The effect of nutrient enrichment in Lake Erie (Beeton, 1969) has changed the phytoplankton composition as well as increasing the concentration markedly. Zooplankton species, formerly negligible, have become abundant. The bottom is enriched with organic matter and the water just above it is often depleted in oxygen, both factors causing changes in the bottom fauna. Lake Erie has produced and continues to produce about twenty thousand tons of fish per year, so that there is no marked increase in total fish yield. However, the species composition of the catch has changed markedly, with the greatest changes in the years 1940–1965. Four species have almost disappeared and others have become abundant in the fishery.

The same general features are found in other lakes. In Lake Washington where yearly nutrient and chlorophyll cycles have been followed in detail (Edmonson, 1969), it is interesting to note that when eutrophication was only moderate (in 1950) there was a spring phytoplankton outburst in May which rapidly decreased during June and July, following a pattern similar to that experienced in the sea or simulated in the model. In the years of maximum eutrophication (1963–1964) there appeared to be no decrease of this kind and the concentration of phosphorus in the water in March about matched that in the phytoplankton three months later. Thus, from 1950 to 1963–1964 the winter level of phosphorus increased by a factor of four, but the summer values of chlorophyll increased by about fifteen times. This is a measure of the breakdown of control of the phytoplankton populations.

Brooks (1969) has pointed out that changes in phytoplankton species composition may be as important to the zooplankton as the increase in abundance, since changes in size, particularly the dominance of larger blue-green algae, may be inappropriate for the size of copepods originally present. Alterations in zooplankton composition would then be a response to the phytoplankton changes, but it appears that in many cases of high eutrophication such as Lake Erie and possibly Lake Washington there is decreased ability to cope with the rising plant biomass. In consequence, much of this material is not utilized immediately but ends on the bottom and leads to the deoxygenation which is an extra factor in altering the rest of the food chain.

This outline of the consequences of nutrient enrichment in large lakes shows clearly that effects acting on the lower end of the food chain alter the rest of the system drastically. In the Baltic (Fonselius, 1970) the effects of nutrient enrichment are becoming apparent, and the deoxygenation in deep water could be in part the consequence of lack of utilization of plant material in the upper layers as well as depending on the renewal of deep water from the North Sea.

Other evidence for the effects in the sea of nutrient changes are more tenuous. Red tides caused by very large dinoflagellate blooms have been associated with high nutrients, among other factors. Widespread blooms of *Gonyaulax tamarensis* occurred off the east coast of Britain in 1968 (Wood et al., 1968) and led to deaths of seabirds and to paralytic shellfish poisoning, indicating the effects on the food chain. In the northern North Sea, chlorophyll concentrations greater than 50 mg/m^3 were found where the normal maximum is 5 mg/m^3. Most, if not all, of these blooms occur through natural causes; still, they provide an index of the explosive potential of the phytoplankton when an unusual species occurs which the zooplankton cannot control.

In the model the effect of a simple increase in nutrients is to increase the zooplankton population rather than the phytoplankton. However, if large increases in nutrients were to alter the species composition of the plants, particularly by producing dominance of single previously scarce species, then one could expect both a lower rate of "volume swept clear" and also a disappearance of the threshold response. In this case the theoretical response would be closer to that based on the simple set of assumptions where the only practical solution is a large plant and a small herbivore population. Theoretically, therefore, the changes in species composition are likely to be more important than the increasing plant production in giving the effects typically associated with eutrophication. This alteration, where particular plant species become excessively dominant, is toward a lower "diversity" and superficially would support the idea that the low diversity often associated with polluted conditions is the explanation for the structural alteration of such systems. The connection is only meaningful, however, if one predicates patterns of behavior of the kind used for zooplankton in the model.

The effects of fishing and eutrophication can be used as a background for the consideration of other ways in which we can alter marine food webs in a manner deleterious to ourselves. We are concerned with two quite different types of consequence: those which affect us, either through our food supply or our amenity, without really altering major parts of the marine environment; and those where the breakdown of a sufficiently large part of the structure decreases the advantages we can derive from the sea. At present overfishing is still the most obvious way by which we alter marine resources, but the effect seems confined to the resource itself. Eutrophication is a potential example of the breakdown of a whole system.

The problems raised by these two possibilities, and the resultant actions we should take, are most acute for those widespread low-level contaminants such as DDT that show "biological amplification" as they pass through a food chain. Do these toxins cause any alterations in the whole ecosystem or only in top predators like seabirds, seals, or man? Since there is considerable pressure to ensure that these organisms do not suffer ill effects, the first, simplest, and most extreme hypothesis is to assume that the rest of the system will be relatively unaffected. While reproductive failure is the main danger, there may be synergistic effects of toxins in combination with natural events. There is some evidence that the mass mortalities of seabirds in the northern Irish Sea in 1969 resulted mainly from starvation and that the high PCB concentrations found in some birds were an additive factor rather than the sole cause of death (NERC, 1971).

For fish at a lower trophic level, the effects of contaminants which show amplification up the food chain would be expected to be smaller. However, the combination of this type of toxin with heavy and sustained exploitation might have more serious consequences for recruitment of commercial species than the occasional natural catastrophe. Even so, since large changes in fisheries do not appear to upset the lower trophic levels, the main impact is likely to be on man rather than on the plankton.

A priori, the direct effects of cumulative toxins on the plant-herbivore component should be much less, for the body burdens will be much lower. At present almost nothing is known of the effects within these groups. Experimental work on oceanic phytoplankton (Menzel, Anderson, and Randtke, 1970) in relation to DDT concentration in seawater showed significant decreases in photosynthesis only at levels above saturation (1 ppb). For zooplankton, however, Menzel and his co-workers (reported in MIT, 1970) have evidence that the offspring of copepods exposed to concentrations of less than 10×10^{-12} DDT do not reach maturity. Although this very low concentration may still be well above that in the open sea, it is an indication that the idea of increasing sensitivity with increasing trophic level does not necessarily hold true.

A report (MIT, 1970) discussing the use of pesticides for insect control on cultivated land points out that such applications destroy not only the herbivorous mites, but also the predatory species. The latter recover more slowly than the former, so that the use of pesticides becomes "addictive." This is given as an illustration of predator control of herbivores and of the importance of conserving predators in this type of farming. Unlike farming, in the sea we harvest predators; yet the conclusion would be that the main problem is the effect of added stress of pollutants on our rate of harvesting rather than on the rest of the system. Although the idea that the predator populations are an index of the "health" of the whole community may be applicable on land, it will not be so advantageous in the study of marine systems.

It is curious, since it is a reversal of the situation on land, that an increase in primary production should be the main method of altering aquatic systems. This has occurred principally in freshwater lakes, and it is difficult to think of growth-promoting effects which could act directly on the phytoplankton in the open sea. Eutrophication is likely to be a problem only in coastal or semienclosed areas near large industrial and populated centers. If other substances cause a moderate *decrease* in the rate of plant production for a given rate of nutrient supply, the model implies that this could be accommodated by relatively small shifts to a new balance of nutrient concentration, plant growth rate, and herbivore grazing. This depends on the continued existence of appropriate herbivore feeding patterns acting as a control on the plants. A priori these patterns are less likely to be affected by a decrease in abundance of one or two plant species than by a marked increase in previously minor components. In either case it is the response of the herbivores which are most important. These responses need to be studied in terms of feeding behavior and reproduction rather than as mortality rates or even as simple metabolic changes. The highly sensitive response to DDT shown by Menzel and his associates demonstrates the need for experiments on the effects over several breeding cycles.

In relatively productive waters approximately 30 percent of plant material produced passes out of the upper layers after ingestion by zooplankton. On the continental shelf this is the source of energy for the benthic fauna and thus for an important part of our fishery yield. At a guess, 30 percent of any toxin entering the system through the phytoplankton will reach the bottom. I have mentioned how little we know of the first stages in processing this material, particularly the quantitative importance of bacteria. Yet these stages are crucial for our understanding of the fallout of toxins in coastal areas before they reach the open seas, of their retention time in the bottom of these areas, and of their eventual appearance in demersal fish. It is known that microorganisms can transform mercury into the highly toxic form, methyl mercury (Jensen and Jernelov, 1969), but an appreciation of the relevance of this transformation depends on an understanding of the dynamics of this part of the web.

I have pointed out that one can expect control or density-dependent patterns to be widespread in any food web, as their occurrence anywhere in the web will tend to improve its response to environmental fluctuations. It is possible that any of these patterns can be altered by very low concentrations of some existing or future pollutant. Such interference with functions will "harm" the system, and we cannot hope to predict where in the web—or how—such harm will occur. Obviously it is desirable to investigate all possible effects at all trophic levels.

However, there is often a need to restrict work, or even to speculate on possible consequences with only limited information or time. The differences between marine and terrestrial systems, which has been one of my themes, would suggest that our selection of problems in the sea must be almost exactly opposite to those on land. We monitor the body burden of toxins we get from terrestrial plants and herbivores, but worry about the effects on the system of destruction of predators. In the sea we are now beginning to estimate the amounts of several pollutants in the predatory fish; still, any effect on the system may be most critical at intermediate levels, principally the pelagic herbivores, but also the benthic microorganisms.

The bias of this exposition has been toward a theoretical approach. In itself such an approach, although it can help to display problems more clearly, does not provide answers. It is only part of a process whereby previous field observations and experiments are brought together and their consequences interpreted in a way which suggests what further observations and experiments will provide the most critical tests of one's ideas. The need for experimental work is obvious, and it is possible that the main links between terrestrial and marine ecology may derive from similarities in their results.

Perhaps the main problem is the collection of observations in the open sea. The difficulties in obtaining sufficient detail, while partly logistic, are also a question of our insight into an essentially obscure environment. In the past the main problems of management in the sea have centered on certain carnivores, the commercially exploited fish stocks—for which, usually, we have more information than for any part of the food web and which are comparable to the predators on land. If these speculations have any practical consequence, they suggest that in the future our critical problems may occur at lower trophic levels and that we must devote an equal effort to studying their structure and dynamics.

Glossary, References, Index

Glossary

BENTHOS. A general term for the communities found on the seabed. It is often divided into three categories: meiobenthos—the smaller fraction passing through an 0.5 mm sieve; macrobenthos—organisms caught by grabs or dredges and retained by an 0.5 mm sieve; and epibenthos organisms (usually larger) living on, rather than in, the seabed.

CHLOROPHYLL. Pigments associated with photosynthesis in plants, of peculiar importance in the sea because they can be measured at extremely low concentrations. Normally it is not possible to measure directly the biomass of phytoplankton; thus one pigment, chlorophyll *a*, is widely used as an index of the plant material in the sea.

CONTINENTAL SHELF. The areas near land where the depth is less than 200 meters. Beyond this limit the depth increases rapidly, usually to more than 2,000 meters. Demersal fisheries are confined to the shelf, where the benthic production is much larger than in deep water.

COPEPODS. Crustacea which are the dominant group in the zooplankton. Best known are the *Calanus* species reaching, as adults, about 100 to 500 μg dry weight. The females produce 100 to 1,000 eggs, which then go through five *naupliar* stages followed by five *copepodite* stages before becoming adults (also called Stage VI).

DEMERSAL. Fish species which live on the seabed and feed on other organisms living there.

DIATOMS. Unicellular plants with a skeleton or capsule made up of two valves which are mainly composed of silica. Normally the dominant phytoplankton in any period of rapid general growth such as the spring bloom.

DIFFUSION. The horizontal turbulence of the sea produces lateral mixing or diffusion of the water similar in some ways to processes of molecular diffusion but on a much larger scale.

EFFICIENCY. There are many ratios which can be used to describe the transfer of energy from a prey to a predator species. There are, however, two main categories: growth (transfer) efficiency referring to processes within the life span of an individual; and ecological (food chain) efficiency where the time span is sufficiently long to include all stages in the life cycles of both prey and predator. Beyond this, one can use for the denominator either the total energy ingested by the predator or only the energy assimilated; or one can consider the relation of a prey to one, or to all, of its predators. In this way many definitions can be generated, and I have tried to define the term each time it is used.

EUPHOTIC ZONE. Photosynthesis by phytoplankton depends on the penetration of light into the sea. The depth to which sufficient light penetrates defines the euphotic zone and can vary from a few meters in turbid inshore waters to 120 meters in the Sargasso Sea.

EUTROPHICATION. The increase in plant growth rate resulting from an increased rate of nutrient supply. Whereas phosphorus is a major problem in fresh water, nitrogen addition appears to be the critical factor in the sea. The addition of nutrients by man is usually called "cultural" eutrophication.

MIXED LAYER. When a thermocline is formed, there is usually a layer above it which wind-induced wave action keeps well mixed. It is in this upper layer that most of the primary production by phytoplankton occurs.

NUTRIENT. Of the many nutrient elements required by phytoplankton a few appear to be limiting. Phosphate and nitrate are considered most important and are most frequently measured. Silicate occupies a special position, since it is essential for diatoms.

PELAGIC. Those members of the food chain living in the water column rather than on the bottom. Particularly applied to fish, such as herring or anchovy, which feed on plankton.

PHYTOPLANKTON. Microscopic algae, generally unicellular, which are found predominantly in the top 100 meters of the sea. Usually 5 to 50 μ in diameter.

PURSE-SEINE. A method of fishing for pelagic species (such as herring or anchovy) where a concentration of fish is enclosed by a net. The high efficiency of this method can cause concern about the reduction of stocks to such low levels that recruitment may be affected.

RECRUITMENT. In the study of fish populations, the numbers of fish reaching a size, or age, at which they are caught by commercial fishing gear. For species which reproduce annually, it provides an index of the success of each new brood.

SALINITY. The salt content of the sea, around 3.0 to 3.5 percent, is regularly measured, partly because it provides an indication of the origin (coastal vs oceanic) of the water mass being studied, partly because vertical changes in salinity, like similar changes in temperature, can cause vertical stability.

SPRING BLOOM. In temperate waters, the formation of a thermocline in the spring gives an upper layer with relatively high nutrient content. The ensuing rapid growth of phytoplankton before the nutrients are used up usually produces a high-standing crop of phytoplankton referred to as the spring bloom.

STOCHASTIC. A mathematical technique used to assess the effects of random variations on some theoretically defined process. For example, random variations can be introduced in a biological model to simulate the effects of natural fluctuations in the physical environment.

THERMOCLINE. Heating of the surface waters of the sea normally produces an upper, warmer layer separated from the deeper water by a thermocline.

The warmer water is less dense, and mixing between the upper and lower zones is inhibited, providing vertical stability and a partial isolation of this upper water. In the tropics this shallow thermocline is always present, but in temperate waters it forms in the spring and breaks down in the autumn.

UPWELLING. In certain coastal sea areas the prevailing winds tend to displace the surface layers offshore, and these are replaced by deeper, nutrient-rich water. The continuous supply of nutrients gives high rates of primary production in such areas as Peru and West Africa. The phytoplankton are grazed on by fish such as anchovy, and it is this straight conversion from plants to fish, as much as the high basic production, which is responsible for the high fish yields in these areas.

ZOOPLANKTON. The smaller animals which occur mainly in the upper layers of the sea. They are caught with silk or nylon nets having mesh sizes normally ranging from 0.1 to 1.0 mm. The majority of animals caught in this way are herbivores, typically copepods, but the zooplankton also include smaller invertebrate carnivores.

References

Adams, J.A., and J.H. Steele. 1966. Shipboard experiments on the feeding of *Calanus finmarchicus* (Gunnerus). In *Some contemporary studies in marine science*, H. Barnes, ed. Allen & Unwin, London, pp. 19–35.

Andrews, P., and P.J. LeB. Williams. 1971. Heterotrophic utilisation of dissolved organic compounds in the sea. III. Measurement of the oxidation rates and concentrations of glucose and amino acids in sea water. *J. Mar. Biol. Assn. U.K.*, 51:111–126.

Anilov, I.K., A.A. Elizarov, I.P. Kanaeva, and G.N. Lavrov. 1969. Bio-oceanological premises of the search for krill. *Trudy VNIRO* 66:246–248.

Bainbridge, V. 1963. Continuous plankton records: contribution towards a plankton atlas of the North Atlantic and North Sea. VIII. Chaetognatha. *Bull. Mar. Ecol.* 4:40–51.

Bartlett, M.S. 1957. Competitive and predatory biological systems. *Biometrika* 44:29–31.

Beers, J.R. 1966. Studies on the chemical composition of the major zooplankton groups in the Sargasso Sea off Bermuda. *Limnol. Oceanogr.* 11:520–528.

Beeton, A.M. 1969. Changes in the environment and biota of the Great Lakes. In *Eutrophication*, Nat. Acad. Sci., Washington, D.C., pp. 150–187.

Beklemishev, C.W. 1962. Superfluous feeding of marine herbivorous zooplankton. *Rapp. p-v. Réun. Cons. perm. int. Explor. Mer* 153:108–113.

Blaxter, K.L. 1962. *The energy metabolism of ruminants.* Hutchinson, London.

Bowden, K.F. 1960. Turbulence. In *The sea*, vol. 1, M.N. Hill, ed. Wiley-Interscience, London, pp. 802–825.

Brooks, J.L. 1969. Eutrophication and changes in the composition of zooplankton. In *Eutrophication*, Nat. Acad. Sci., Washington, D.C., pp. 236–255.

Butler, E.I., E.D.S. Corner, and S.M. Marshall. 1969. On the nutrition and metabolism of zooplankton. VI. Feeding efficiency of *Calanus* in terms of nitrogen and phosphorus. *J. Mar. Biol. Assn. U.K.*, 49:977–1002.

——————1970. On the nutrition and metabolism of zooplankton. VII. Seasonal survey of nitrogen and phosphorus excretion of *Calanus* in the Clyde Sea area. *J. Mar. Biol. Assn. U.K.* 50:525–560.

References

Clark, L.R., P.W. Geier, R.D. Hughes, and R.F. Morris. 1967. *The ecology of insect populations in theory and practice.* Methuen, London.

Colebrook, J.M., and G.A. Robinson. 1961. The seasonal cycle of the plankton in the North Sea and the north-eastern Atlantic. *J. Cons. Int. Explor. Mer* 26:156–165.

Conover, R.J. 1962. Metabolism and growth in *Calanus hyperboreus* in relation to its life cycle. *Rapp. p-v. Réun. Cons. perm. int. Explor. Mer* 153:190–197.

——1966. Factors affecting the assimilation of organic matter by zooplankton and the question of superfluous feeding. *Limnol. Oceanogr.* 11:346–354.

Corner, E.D.S., and C.B. Cowey. 1968. Biochemical studies on the production of marine zooplankton. *Biol. Rev.* 43:393–426.

Crisp, D.J. 1964. Introduction to *Grazing in terrestrial and marine environments,* D.J. Crisp, ed. Blackwell, Oxford, pp. xi–xvi.

Cushing, D.H. 1959. On the nature of production in the sea. *Fish. Invest., London,* ser. 2, vol. 22, no. 6.

——1961. On the failure of the Plymouth herring fishery. *J. Mar. Biol. Assn. U.K.* 41:799–816.

——1969. *Upwelling and fish production.* FAO Fish. Tech. Pap., no. 84.

——and D.S. Tungate. 1963. Studies on a *Calanus* patch. I. The identification of a *Calanus* patch. *J. Mar. Biol. Assn. U.K.* 43:327–337.

Dickson, R., and A.J. Lee. 1969. Atmospheric and marine climate fluctuations in the North Atlantic region. *Prog. Oceanogr.* 5:55–65.

Dunbar, M.J. 1968. *Ecological development in polar regions.* Prentice-Hall, London.

Duursma, E.K. 1961. Dissolved organic carbon, nitrogen and phosphorous in the sea. *Neth. J. Sea. Res.* 1:1–148.

Edmondson, W.T. 1969. Eutrophication in North America. In *Eutrophication,* Nat. Acad. Sci., Washington, D.C., pp. 124–149.

Edwards, C.A., and G.W. Heath. 1963. The role of soil animals in the breakdown of leaf material. In *Soil organisms,* J. Doeksen and J. van der Drift, eds. North-Holland Publishing Company, Amsterdam.

Edwards, R.R.C., D.M. Finlayson, and J.H. Steele. 1969. The ecology of O-group plaice and common dabs in Loch Ewe. II. Experimental studies of metabolism. *J. Exp. Mar. Biol. Ecol.* 3:1–17.

Engelmann, M.D. 1966. Energetics, terrestrial field studies and animal productivity. In *Advances in ecological research,* J.B. Cragg, ed., vol. 3. Academic Press, New York, pp. 73–115.

Eppley, R.W., and W.H. Thomas. 1969. Comparison of half-saturation constants for growth and nitrate uptake of marine phytoplankton. *J. Physiol.* 5:375–379.

——J.N. Rogers, and J.J. McCarthy. 1969. Half-saturation constants for uptake of nitrate and ammonium by marine phytoplankton. *Limnol. Oceanogr.* 14:912–920.

Fonselius, S.H. 1970. On the stagnation and recent turnover of the water in the Baltic. *Tellus* 22:533–544.

Fraser, J.H. 1970. The ecology of the ctenophore *Pleurobrachia pileus* in

Scottish waters. *J. Cons. Int. Explor. Mer* 33:149–168.
Gauld, D.T. 1951. The grazing rate of planktonic copepods. *J. Mar. Biol. Assn. U.K.* 29:695–706.
Gerbach, S.A. 1971. On the importance of marine meiofauna for benthos communities. *Oecologia* 6:176–190.
Gibb, J.A. 1966. Tit predation and the abundance of *Ernarmonia conicolana* (Heyl.) on Weeting Heath, Norfolk, 1962–63. *J. Anim. Ecol.* 35:43–54.
Glover, R.S., G.A. Robinson, and J.M. Colebrook. 1970. Plankton in the North Atlantic—an example of the problems of analysing variability in the environment. *FAO Tech. Conf. on Mar. Poll.*, FIR:MP/70/E-55.
Gulland, J.A. 1970. Food chain studies and some problems in world fisheries. In *Marine food chains,* J.H. Steele, ed. Oliver & Boyd, Edinburgh, pp. 296–318.
Hairston, N.G., F.E. Smith, and L.B. Slobodkin. 1960. Community structure, population control and competition. *Amer. Natur.* 94:421–425.
Hardy, A.C. 1924. The herring in relation to its animate environment. I. The food and feeding habits of the herring with special reference to the east coast of England. *Fish. Invest., Lond.*, ser. 2, vol. 7, no. 3.
——and E.R. Gunther. 1935. The plankton of the South Georgia whaling ground and adjacent waters, 1926–7. *Discovery Rep.* 11:1–456.
Hargrave, B.T. 1970. The effect of a deposit-feeding amphipod on the metabolism of benthic microflora. *Limnol. Oceanogr.* 15:21–30.
Harvey, H.W. 1937. Note on selective feeding by *Calanus. J. Mar. Biol. Assn., U.K.* 22:97–100.
Holling, C.S. 1965. The functional response of predators to prey density and its role in mimicry and population regulation. *Mem. ent. Soc. Can.* 45:5–60.
Huffaker, C.B. 1958. Experimental studies on predation: dispersion factors and predator-prey oscillations. *Hilgardia* 27:343–383.
Hutchinson, G.E. 1948. Circular causal systems in ecology. *Ann. N.Y. Acad. Sci.* 50:221–246.
International Council for the Exploration of the Sea. 1969. Report of the working group on assessment of demersal species in the North Sea. *Coop. Res. Rept., ser. A, no. 9, Int. Council Explor. Sea.*
——1970. Liaison Committee Report. *Int. Counc. Explor. Mer Coop. Res. Rept, ser. B.*
Ivlev, V.S. 1961. *Experimental ecology of the feeding of fishes*, D. Scott, trans. Yale University Press, New Haven.
Jensen, S., and A. Jernelov. 1969. Biological methylation of mercury in aquatic organisms. *Nature* 223:753–754.
Johannes, R.E., and M. Satomi. 1966. Composition and nutritive value of faecal pellets of a marine crustacean. *Limnol. Oceanogr.* 11:191–197.
Johnson, C.G. 1969. *Migration and dispersal of insects by flight.* Methuen, London.
Jones, R. 1954. The food of the whiting, and a comparison with that of the haddock. *Mar. Res.*, no. 2.
——1973. Density dependent regulation of the numbers of cod and haddock. *Rapp. p-v. Réun. Cons. perm. int. Explor. Mer* (in press).
Joseph, J. and H. Sendner. 1958. Über die horizontale Diffusion in Meere. *Dt.*

hydrogr. Z. 11:49–77.
Kierstead, H., and L.B. Slobodkin. 1953. The size of water masses containing plankton blooms. *J. Mar. Res.* 12:141–147.
Kneese, A.V., R.U. Ayres, and R.C. d'Arge. 1970. *Economics and the environment.* Johns Hopkins Press, Baltimore.
Lack, D. 1966. *Population studies of birds.* Oxford University Press, London.
Lasker, R. 1970. Utilization of zooplankton energy by a Pacific sardine population in the California current. In *Marine food chains,* J.H. Steele, ed. Oliver & Boyd, Edinburgh, pp. 265–284.
McAllister, C.D. 1970. Zooplankton rations, phytoplankton mortality and the estimation of marine production. In *Marine food chains*, J.H. Steele, ed. Oliver & Boyd, Edinburgh, pp. 419–457.
MacArthur, R.H. 1955. Fluctuations of animal populations, and a measure of community stability. *Ecology* 36:533–536.
——1957. The relative abundance of bird species. *Proc. Amer. Acad. Sci.* 43:293–295.
Macfadyen, A. 1961. Metabolism of soil invertebrates in relation to soil fertility. *Ann. Appl. Biol.* 49:215–218.
——1964. Energy flow in ecosystems and its exploitation by grazing. In *Grazing in terrestrial and marine environments*, D.J. Crisp, ed., pp. 3–24.
McIntyre, A.D. 1961. Quantitative differences in the fauna of boreal mud associations. *J. Mar. Biol. Assn. U.K.* 41:599–616.
——1969. Ecology of marine meiobenthos. *Biol. Rev.* 44:245–290.
—— A.L.S. Munro, and J.H. Steele. 1970. Energy flow in a sand ecosystem. In *Marine food chains*, J.H. Steele, ed. Oliver & Boyd, Edinburgh, pp. 19–31.
MacIsaac, J.J., and R.C. Dugdale. 1969. The kinetics of nitrate and ammonia uptake by natural populations of marine phytoplankton. *Deep-Sea Res.* 16:45–58.
McLaren, I.A. 1963. Effects of temperature on the growth of zooplankton and the adaptive value of vertical migration. *J. Fish. Res. Bd. Can.* 20:685–727.
McNeill, S., and J.H. Lawton. 1970. Annual production and respiration in animal populations. *Nature* 225:472–474.
McQueen, D.J. 1970. Grazing rates and food selection in *Diaptomus oregonensis* (Copepoda) from Marion Lake, British Columbia. *J. Fish. Res. Bd. Can.* 27:13–20.
Marshall, S.M., and A.P. Orr. 1955. *The biology of a marine copepod.* Oliver & Boyd, Edinburgh.
————1958. On the biology of *Calanus finmarchicus.* X. Seasonal changes in oxygen consumption. *J. Mar. Biol. Assn. U.K.* 37:459–472.
Marti, J.J. 1967. Biological resources of the Scotia Sea and neighbouring regions. Akad. Nauk. S.S.R., Moscow, pp. 140–145 (unpublished translation).
Massachusetts Institute of Technology 1970. *Man's impact on the global environment.* MIT Press, Cambridge, Mass.
May, R.M. 1971. Stability in multi-species community models. *Math. Biosci.* 12:59–79.
Menzel, D.W. 1967. Particulate organic carbon in the deep sea. *Deep-Sea Res.* 14:229–238.
——and J.H. Ryther. 1960. The annual cycle of primary production in the

Sargasso Sea off Bermuda. *Deep-Sea Res.* 6:351–367.
——— ———1961. Zooplankton in the Sargasso Sea off Bermuda and its relation to organic production. *J. Cons. Int. Explor. Mer* 26:250–258.
——, J. Anderson, and A. Randtke. 1970. Marine phytoplankton vary their response to chlorinated hydrocarbons. *Science* 167:1724–1726.
Mullin, M.M. 1963. Some factors affecting the feeding of marine copepods of the genus *Calanus. Limnol. Oceanogr.* 8:239–250.
——and E.R. Brooks. 1970a. Growth and metabolism of two planktonic copepods as influenced by temperature and type of food. In *Marine food chains,* J.H. Steele, ed. Oliver & Boyd, Edinburgh, pp. 74–95.
——— ———1970b. The effect of concentration of food on body weight, cumulative ingestion, and rate of growth of the marine copepod *Calanus helgolandicus. Limnol. Oceanogr.* 15:748–755.
Murdoch, W.W. 1969. Switching in general predators: experiments on predator specificity and stability of prey populations. *Ecol. Monog.* 39:335–354.
Murphy, G.I. 1968. Patterns in life history and the environment. *Amer. Nat.* 102:391–404.
National Environmental Research Council. 1971. The sea bird wreck in the Irish Sea, autumn 1969. *NERC Publ.* ser. C, no. 4.
Newell, R. 1965. The role of detritus in the nutrition of two marine deposit feeders, the prosobranch *Hydrobia ulvae* and the bivalve *Macoma baltica. Proc. Zool. Soc. Lond.* 144:25–45.
Odum, E.P. 1971. *Fundamentals of ecology.* Saunders, Philadelphia.
Paffenhofer, G.A. 1971. Grazing and ingestion rates of nauplii, copepodids and adults of the marine planktonic copepod *Calanus helgolandicus. Mar. Biol.* 11:286–298.
——and J.D.H. Strickland. 1970. A note on the feeding of *Calanus helgolandicus* on detritus. *Mar. Biol.* 5:97–99.
Paine, R.T. 1969. A note on trophic complexity and community stability. *Amer. Nat.* 103:91–93.
Parsons, T.R., and H. Seki. 1970. Importance and general implications of organic matter in aquatic environments. In *Organic matter in natural waters,* D.W. Hood, ed. Inst. Mar. Sci., U. of Alaska, pub. no. 1, pp. 1–27.
——and R.J. LeBresseur, J.D. Fulton, and O.D. Kennedy. 1969. Production studies in the Strait of Georgia. II. Secondary production under the Fraser River plume, February to May, 1967. *J. Exp. Mar. Biol. Ecol.* 3:39–50.
——K. Stephens, and J.D.H. Strickland. 1961. On the chemical composition of eleven species of marine phytoplankton. *J. Fish. Res. Bd. Can.* 18:1001–1016.
Petipa, T.S., E.V. Pavlova, and G.N. Mironov. 1970. The food web structure, utilization and transport of energy by trophic levels in the planktonic communities. In *Marine food chains*, J.H. Steele, ed. Oliver & Boyd, Edinburgh, pp. 143–167.
Phillipson, J. 1966. *Ecological energetics.* Edward Arnold, London.
Pielou, E.C. 1969. *An introduction to mathematical ecology.* Wiley-Interscience, London.

Pitelka, F.A. 1964. The nutrient recovery hypothesis for arctic microtine cycles. I. Introduction. In *Grazing in terrestrial and marine environments*, D.J. Crisp, ed. Blackwell, Oxford, pp. 55–56.

Platt, T. 1972. Local phytoplankton abundance and turbulence. *Deep-sea Res.* 19:183–188.

Rae, B.B. 1967. The food of the cod in the North Sea and on west of Scotland grounds. *Mar. Res.*, no. 1.

Redfield, A.C., B.H. Ketchum, and F.A. Richards. 1963. The influence of organisms on the composition of sea water. In *The sea*, vol. 2, M.N. Hill, ed. Wiley-Interscience, London, pp. 26–77.

Reeve, M.R. 1963. Growth efficiency in *Artemia* under laboratory conditions. *Biol. Bull. Mar. Biol. Lab. Woods Hole* 125:133–145.

——1970. The biology of Chaetognatha. I. Quantitative aspects of growth and reproduction in *Sagitta hispida*. In *Marine food chains*, J.H. Steele, ed. Oliver & Boyd, Edinburgh, pp. 168–192.

Richman, S., and J.N. Rogers. 1969. The feeding of *Calanus helgolandicus* on synchronously growing populations of the marine diatom *Ditylum brightwellii*. *Limnol. Oceanogr.* 14:701–709.

Riley, G.A. 1963. Organic aggregates in sea water and the dynamics of their formation and utilization. *Limnol. Oceanogr.* 8:372–381.

Rosenthal, H., and G. Hempel. 1970. Experimental studies in feeding and food requirements of herring larvae (*Clupea harengus* L.). In *Marine food chains*, J.H. Steele, ed. Oliver & Boyd, Edinburgh, pp. 344–364.

Russell, F.S., A.J. Southward, G.T. Boalch, and E.I. Butler. 1971. Changes in biological conditions in the English Channel off Plymouth during the last half century. *Nature* 234:468–470.

Ryther, J.H. 1969. Relationship of photosynthesis to fish production in the sea. *Science* 166:72–76.

Sanders, H.L. 1956. Oceanography of Long Island Sound, 1952–54. 10. Marine bottom communities. *Bull. Bingham Oceanogr. Coll.* 15:345–414.

——1968. Marine benthic diversity: a comparative study. *Amer. Nat.* 102:243–282.

——1969. Benthic diversity and the stability-time hypothesis. *Brookhaven Symp. in Biol.* 22:71–81.

Savage, R.E. 1937. The food of the North Sea herring, 1930–1934. *Fish. Invest., Lond.*, ser. 2, vol. 15, no. 5.

Saville, A. 1956. Eggs and larvae of haddock (*Gadus aeglefinus* L.) at Faroe. *Mar. Res.*, no. 4.

Scudo, F.M. 1971. Vito Volterra and theoretical ecology. *Theor. Pop. Biol.* 2:1–23.

Simberloff, D.S., and E.O. Wilson. 1969. Experimental zoogeography of islands. The colonisation of empty islands. *Ecology* 50:278–295.

Slobodkin, L.B. 1961. *Growth and regulation of animal populations.* Holt, Rinehart & Winston, New York.

——and H.L. Sanders. 1969. On the contribution of environmental predictability to species diversity. *Brookhaven Symp. in Biol.* 22:82–95.

Smayda, T.J. 1969. Some measurements of the sinking rate of faecal pellets. *Limnol. Oceanogr.* 14:621–625.

Steele, J.H. 1958. Plant production in the northern North Sea. *Mar. Res.*, no. 7.

——1961. Primary production. In *Oceanography*, Mary Sears, ed. Amer. Assn. Adv. Sci., publ. no. 67, pp. 519–538.

——1965a. Notes on some theoretical problems in production ecology. *Mem. Ist. Ital. Idrobiol.*, 18 suppl.: 383–398.

——1965b. Some problems in the study of marine resources. *Int. Comm. N.W. Atlantic Fish.*, spec. pub. 6:463–476.

——and I.E. Baird, 1961. Relations between primary production, chlorophyll and particulate carbon. *Limnol. Oceanogr.* 6:68–78.

————1962. Further relations between primary production, chlorophyll and particulate carbon. *Limnol. Oceanogr.* 7:42–47.

————1965. The chlorophyll *a* content of particulate organic matter in the northern North Sea. *Limnol. Oceanogr.* 10:261–267.

————1972. Sedimentation of organic matter in a Scottish sea loch. *Mem. Ist. Ital. Idrobiol.*, 29 suppl., 73–88.

——and D.W. Menzel. 1962. Conditions for maximum primary production in the mixed layer. *Deep-Sea Res.* 9:39–49.

——A.D. McIntyre, R.R.C. Edwards, and A. Trevallion. 1970. Interrelations of a young plaice population with its invertebrate food supply. In *Animal populations in relation to their food resources*, A. Watson, ed. Blackwell, Oxford, pp. 373–387.

Steemann Nielsen, E. 1952. The use of radio-active carbon (C^{14}) for measuring organic production in the sea. *J. Cons. int. Explor. Mer* 18:117–140.

Stommel, H., and A.B. Aarons. 1960. On the abyssal circulation of the world ocean. II. An idealised model of the circulation pattern and amplitude in ocean basins. *Deep-Sea Res.* 6:217–233.

Tinbergen, L. 1960. The natural control of insects in pine woods. I. Factors influencing the intensity of predation by song-birds. *Arch. néerl. Zool.* 13:265–335.

Vlymen, W.J. 1970. Energy expenditure of swimming copepods. *Limnol. Oceanogr.* 15(3): 348–356.

Wangersky, P.J., and W.J. Cunningham. 1957. Time lag in prey predator population models. *Ecology* 38:136–139.

Watson, A., and R. Moss. 1970. Dominance, spacing behaviour and aggression in relation to population limitation in vertebrates. In *Animal populations in relation to their food resources*, A. Watson, ed. Blackwell, Oxford, pp. 167–220.

Westlake, D.F. 1963. Comparisons of plant productivity. *Biol. Rev.* 38:385–425.

Wiebe, P.H. 1970. Small scale spatial distribution in oceanic zooplankton. *Limnol. Oceanogr.* 15:205–217.

Winberg, G.G. 1956. Intensivnost obmena i pishehevye potrebnosti ryb *Nauchnye Trudy Belorusskovo Gosudarstvennovo Universiteta Imeni*, V.I. Lenina, Minsk, and *Fish. Res. Bd. Can. Trans. Ser.*, no. 194.

Wood, P.C. et al. 1968. Dinoflagellate crop in the North Sea. *Nature* 220:21–27.

Wynne-Edwards, V.C. 1962. *Animal dispersion in relation to social behaviour.* Oliver & Boyd, Edinburgh.

Index